SHAKESPEARE
and the Mismeasure of Renaissance Man

SHAKESPEARE

and the Mismeasure of Renaissance Man

PAULA BLANK

CORNELL UNIVERSITY PRESS

Ithaca and London

A version of chapter 4 appeared as "Shakespeare's Equalities: Check-ing the Math of *King Lear*" by Paula Blank in *Exemplaria: A Journal of Theory in Medieval and Renaissance Studies* 15.2 (Autumn 2003): 471–506. Pegasus Press, P.O. Box 15806, Asheville, NC 28813. Copy-right © 2003.

First published 2006 by Cornell University Press
Printed in the United States of America

Library of Congress Cataloging-in-Publication Data

Blank, Paula, 1959–
   Shakespeare and the mismeasure of Renaissance man / Paula Blank.
     p. cm.
   Includes bibliographical references and index.
   ISBN-13: 978-0-8014-4475-3 (cloth : alk. paper)
   ISBN-10: 0-8014-4475-6 (cloth : alk. paper)
  1. Shakespeare, William, 1564–1616—Criticism and interpretation. 2. Measurement in literature. 3. Anthropometry in literature. I. Title.
  PR3069.M43B57 2007
  822.3'3—dc22

                              2006006244

Cornell University Press strives to use environmentally responsible suppliers and materials to the fullest extent possible in the publishing of its books. Such materials include vegetable-based, low-VOC inks and acid-free papers that are recycled, totally chlorine-free, or partly composed of nonwood fibers. For further information, visit our web-site at www.cornellpress.cornell.edu.

Cloth printing 10 9 8 7 6 5 4 3 2 1

1. We measure from ourselves; and as things are for our use and purpose, so we approve them. Bring a Pear to the Table that is rotten, we cry it down, 'tis naught; but bring a Medlar that is rotten, and 'tis a fine thing, and yet I'll warrant you the Pear thinks as well of itself as the Medlar does.

2. We measure the Excellency of other Men, by some Excellency we conceive to be in ourselves. *Nash* a Poet, poor enough (as Poets us'd to be), seeing an Alderman with his Gold Chain, upon his great Horse, by way of scorn, said to one of his Companions, do you see yon fellow, how goodly, how big he looks; why that fellow cannot make a blank Verse.

3. Nay we measure the goodness of God from ourselves; we measure his Goodness, his Justice, his Wisdom, by something we call Just, Good, or Wise in ourselves; and in so doing, we judge proportionably to the Country-fellow in the Play, who said if he were a King, he would live like a Lord, and have Peas and Bacon every day, and a Whip that cried Slash.

—JOHN SELDEN (1584–1654), "Measure of Things," in *Table-Talk*

CONTENTS

ACKNOWLEDGMENTS

I would like to thank the National Endowment for the Humanities, the National Humanities Center, the Folger Shakespeare Library, and the College of William and Mary for research opportunities that made it possible for me to write this book. An earlier version of chapter 4 appeared as "Shakespeare's Equalities: Checking the Math of *King Lear*," in *Exemplaria: A Journal of Theory in Medieval and Renaissance Studies* 15.2 (2003): 471–506.

I am very grateful for the help of friends and colleagues who commented on portions of the manuscript, especially Paul Aron, Dayton Haskin, Laura Levine, Adam Potkay, Talbot J. Taylor, and Peter DeSa Wiggins. Many thanks to James Baron and William Hutton for their help with classical materials and translations. I was very fortunate in the two anonymous readers who critiqued the book for Cornell; I have tried to address their remarkably helpful questions and comments as fully as possible. I am also grateful to have benefited from the expertise and support of my editors, Bernhard Kendler and Roger Haydon, and the talents of manuscript editor Karen Hwa. Special thanks to Roland Greene and to Barbara K. Lewalski; as I hope they know, they have my deepest respect, affection, and admiration. This book could not have been fully imagined without Sean Keilen, who provided me with key passages in primary texts, translations, new ways of thinking about my materials, a standard for eloquence, and the pattern for a most extraordinary friendship. Finally, my thanks to Jae Aron, who proves to me that some experiences in life really are beyond measure.

SHAKESPEARE
and the Mismeasure of Renaissance Man

INTRODUCTION

Man is the measure of all things.
—PROTAGORAS, fifth century BC (as cited by Plato)

[W]e're not yet prepared to concede . . . that every man is the
measure of all things, if he isn't an intelligent person. . . . One
person is wiser than another, and it's that sort of person who's a
measure; whereas someone with no knowledge, like me, is in
no way bound to be a measure.—SOCRATES, Plato's *Theaetetus*

FROM ITS FIRST KNOWN CITATION, in Plato's *Theaetetus*,
through the Renaissance and beyond, the Greek philosopher Protagoras's famous dictum, "Man is the measure of all things," has been persist-
ently beset with questions, qualifications, and critique. As it is generally
interpreted, Protagoras's *homo mensura*[1]—the man-measure—determines
"truth" or "reality" according to the scope and scale of human cog-
nizance and capacity. Yet if Protagoras believed that every man, any man,
might provide a fit frame of reference for knowledge, Socrates, for one,
considered the contingencies and limitations of what particular people
bring to the problem of perceiving the world around them, exposing
man—or rather *men and women*—as divergent, uncertain, and variable
measures of all things. Protagoras's apparent emphasis on the nature of
the measure, rather than on the nature of the object of measurement,
as the key to understanding, opened the way for centuries of polemic
about the effect of our technologies of knowing (including our own cognitive faculties) on what we think we know. More broadly, Protagoras's
dictum established the discourse of "measurement" as a means of articu-

---

[1] Here and throughout this book, I follow the long-standing scholarly tradition of
speaking of Protagoras's formulation in its Latin translation. Plato, *Theaetetus*, trans.
John McDowell (Oxford: Clarendon Press, 1973), 183c.

lating what mind performs over matter, a way of talking about how we assess ourselves and our world.

Shakespeare inherited the ancient idea of measurement as a way of representing human apprehension, and the terms of Renaissance measurement systems—scales and spans, squares and levels, ratings and rules—provide his poems and plays with key metaphors of knowing. This book is about Shakespeare's rhetoric of measurement, especially, why his dramatic characters and poetic personae regularly speak of numbering, weighing, or measuring one another. From the poet's constant complaints in the *Sonnets* about faulty estimations, "Thyself thou gav'st, thy own worth then not knowing," to Shylock's problem of gauging an "equal pound" of his nemesis, Antonio, in *The Merchant of Venice*, and from the potentially fatal application of Angelo's "strait" rule in *Measure for Measure* to the king's failed attempt to "weigh equalities," or properly apportion goods to human merit, at the start of *King Lear*,[2] Shakespeare's characters continually engage the language of measurement in the judgments they make of themselves and others.

Though long in circulation as tropes for human understanding, such language had a distinctive resonance during the Renaissance. It is a tenet of the history of science that the modern disciplines of astronomy, medicine, music, navigation, physics, and others arose when Renaissance natural philosophers replaced the "qualitative" view of reality dominant among the ancients with a "quantitative" view based, centrally, on procedures of measurement. Alongside the more familiar, nominalist God who created the world with the Word, Renaissance thinkers advanced the mathematical God who operated somewhat differently, according to the words of Solomon's Wisdom, "Thou hast ordered all things in measure, number & weight."[3] The early seventeenth-century philosopher Comenius thus proceeded from the premise that "in the number, weight, and measure of things great secrets are hidden," while Galileo set out the program of early modern science generally: "To measure what is

[2] William Shakespeare, *The Riverside Shakespeare*, 2nd ed., ed. G. Blakemore Evans (Boston: Houghton Mifflin, 1997), sonnet 87.9; *The Merchant of Venice*, 1.3.149; *Measure for Measure*, 2.1.9; 3.2.255–56; *King Lear*, 1.1.5–6. Throughout this volume, citations of Shakespeare's poems and plays refer to this edition and appear parenthetically in the text.

[3] *The Geneva Bible: Facsimile Edition of 1560*, ed. Lloyd E. Berry (Madison: University of Wisconsin Press, 1969), Wisdom of Salomon 11:17. Throughout this volume, biblical citations refer to this edition of the Geneva Bible and appear parenthetically in the text.

measurable and to try to render measurable what is not so as yet."[4] The predominance of "measuring languages" in Shakespeare's works reflects in part what has been described as "the near frenzy to measure everything imaginable," the "pantometry" or "universal measurement" that characterizes the emergence of scientific thought.[5]

Although Shakespeare may have taken some interest in the measure of the physical world—the quantification of matter, time, and space—he was far more concerned with an attendant cultural phenomenon, what we now call *human* measurement, the attempt to quantify personal, social, and political aspects of human experience. There has been a great deal of scholarly work on the importance of the evolution of measurement systems in the history of science, but far less attention has been paid to corollary attempts in the early modern period to "try to render measurable what is not so as yet" within an immaterial human sphere. Shakespeare's investigation of the Renaissance "measure of reality" centers on the consequences of applying principles of measurement to the appraisal of human nature, conceived as human worth.[6] As a poet and a playwright, Shakespeare was especially attuned to the discursive and performative functions and effects of measurement, including the "rules" of language that helped enable decisions about whether two or more people were to be considered "equal," "less than," or "more than" one another.[7] Shakespeare's interest was not so much the science of meas-

---

[4] Quoted in Karel Berka, *Measurement: Its Concepts, Theories, and Problems*, trans. Augustin Riska, Boston Studies in the Philosophy of Science 72 (Dordrecht, Holland: D. Reidel, 1983), 1, 9.

[5] I have borrowed the phrase "measuring languages" from John E. Murdoch, "From Social into Intellectual Factors: An Aspect of the Unitary Character of Medieval Learning," in *The Cultural Context of Medieval Learning*, ed. John E. Murdoch and Edith Sylla (Dordrecht, Holland: D. Reidel, 1975), 271–348. For a discussion of the sixteenth-century coinage "pantometry" and the preoccupation with mensuration in early modern thought generally, see Alfred W. Crosby, *The Measure of Reality: Quantification and Western Society, 1250–1600* (Cambridge: Cambridge University Press, 1997); Joel Kaye, *Economy and Nature in the Fourteenth Century: Money, Market Exchange, and the Emergence of Scientific Thought* (Cambridge: Cambridge University Press, 1998); and Peter Dear, *Discipline and Experience: The Mathematical Way in the Scientific Revolution* (Chicago: University of Chicago Press, 1995).

[6] The phrase "measure of reality" once again comes from the title of Crosby's book. Attempts at what philosophers of science call "extraphysical" measurement go back at least as far as Aristotle, who in book 2 of his *Ethics*, for example, "measures" the "magnitude" of virtue. In chapter 4, I discuss the Renaissance legacy of Aristotle's "measures" of economic and social justice (*Ethics*, book 5).

[7] "Rule" in early modern English refers to the "ruler" (i.e., a measuring stick or line) as well as to "authority." See the *Oxford English Dictionary*, s.v. "rule."

urement as the Renaissance "art" of measurement, especially the range and variety of rhetorical means available for setting up and solving the problem of human value.

The sheer profusion of measuring terms in Shakespeare's poems and plays, and the frequency of their use, is instructive in itself. His works are rich in reference to "measure, number & weight," the key terms of an early modern quantitative and empirical imagination. In subsequent chapters I investigate specific early modern measures, as Shakespeare adjudged them. Chapter 2, for example, concerns the rhetoric of measurement in Renaissance poetics and the application of Shakespeare's poetic "numbers" to figuring the "worth" of his beloved in the *Sonnets*, along with Shakespeare's use of meter to "measure" human feeling. Chapter 3 considers Renaissance anatomical and other "body" measures and Shakespeare's appraisal of race relations, based on "pounds" of flesh or degrees of kinship. Chapter 4 examines mathematical measures as Renaissance writers, including Shakespeare, applied them to problems of human desert, especially, the determination of economic "equalities"; chapter 5 traces the ancient and enduring use of "measurement" to represent the aims and operations of judgment and the law, as witnessed, notably, by Shakespeare's last comedy. As I hope even this brief outline suggests, Shakespeare draws the language of measurement from numerous early modern disciplines; his usage confirms its currency across the range of Renaissance arts and sciences.

In applying the discourse of physical measurement to the domain of human value, however, Shakespeare more often than not aims to reveal how that discourse is, in effect, *mis*applied. His chronic use of "measuring languages" may well reflect an early modern habit, or desire, to judge people according to clear and authorized standards of evaluation. Yet again and again, as later chapters in this book discover, his characters resist quantification even as they affirm a will for the kind of certitude about how to judge people, in relation to one another, that measurement seems to afford. Indeed, Shakespeare suggests that the most significant rhetorical function of measuring terms, their chief persuasive purpose, is precisely to verify and validate appraisals of human value that are, on closer inspection, biased or unsure.

Shakespeare insists, in other words, that "measurement" must be understood as a figure of speech when it is applied to people, that *human* measurement is a metaphorical operation rather than an empirical one. In the process, he shows us something about the operations of figurative language itself, especially tropes such as metaphor that are based on

making comparisons, on forming relative judgments: "Shall I compare
thee to a summer's day? / Thou art *more* lovely and *more* temperate" (sonnet 18.1–2; emphasis added). As is implicit throughout George Puttenham's *Arte of English Poesie* (1589), for example, Renaissance poetics
is based centrally on problems of "numerositie" and "keeping measure" (see chapter 2). Rehearsing classical and contemporary concerns
about the potential dangers and deceptions of "figures and figurative
speeches," Puttenham construes them as measures or rather "mismeasures" of reality, "as if the carpenter before he began to square his timber
would make his [square] crooked."[8] Shakespeare's copious use of metaphor in his works, his expansive wordplay, and his penchant for linguistic
innovation are enough to prove that he did not share a general distrust
or disapproval of rhetorical expression. Recent work on the "interdisci-
plinarity" of Shakespeare's language, the networks of verbal associations
he creates by crossing the boundaries among Renaissance intellectual,
social, or cultural domains, has, however, sometimes misjudged the in-
tents and effects of the playwright's rhetorical "free play."[9] The fact that
Shakespeare's language is marked by a fluidity of associations, that
Shakespeare so creatively taps the metaphorical potential of words by
traversing multiple semantic fields, does not mean that he was insensitive
to any discontinuities, and disjunctions, he discovers in the process.
Shakespeare crosses lexical borders in part to test the limits of interdisci-
plinarity: his rhetoric of measurement is noteworthy insofar as it confirms the *incommensurability* of the quantifying methods of early physical
science, the measure of time, matter, and space, with problems of
"human" knowledge. For Shakespeare, no doubt, figures of speech potentially extend the scope of how we may imagine ourselves or even con-

[8] George Puttenham, *The Arte of English Poesie*, ed. Gladys Doidge Willcock and Alice
Walker (Cambridge: Cambridge University Press, 1970), 154.
[9] Patricia Parker's work on Shakespeare's language, for example, is based on the
premise that his "word play" extends across all contemporary verbal associations; her
method presumes that Shakespeare treated early modern English as one, borderless,
discursive field, without disciplinary or other boundaries. See Parker, *Shakespeare from
the Margins: Language, Culture, Context* (Chicago: University of Chicago Press, 1996).
Mary Thomas Crane, *Shakespeare's Brain: Reading with Cognitive Theory* (Princeton:
Princeton University Press, 2001), accounts for Shakespeare's metaphors and other
linguistic associations via cognitive theory. For all of the insights this approach af-
fords, it tends to deemphasize what is distinctive about Shakespeare's usage among
his contemporaries. In their shared emphasis on the connections that Shakespeare
draws among words, both studies disregard the extent to which he discriminates
among them as well.

ceive new selves altogether. Yet he also shows how certain figures, if inapt or unbefitting, will prove, instead, the mismeasure of man.[10]

To begin to demonstrate how prevalent measuring terms are in Shakespeare's poems and plays, I would like to offer some preliminary examples of how and why Shakespeare's characters use such language in reference to making human judgments. Hermione, accused of adultery by Leontes in the *Winter's Tale*, pleads that the court "[w]ith thoughts so qualified as [their] charities shall best instruct [them], measure [her]" (2.1.113–14). In the *Second Part of Henry IV*, the Chief Justice tells the Prince, "[I]f I be measured rightly, / Your Majesty hath no just cause to hate me" (5.2.65–66). In *Much Ado about Nothing*, Antonio considers the treachery of the men who have accused his niece, Hero, of being a whore, "I know them, yea, / And what they weigh" (5.1.92–93), while Coriolanus sardonically sums up his feelings for the plebeians: "I love them as they weigh" (2.2.74).

Many of Shakespeare's "measures," moreover, are explicitly comparative: they set up, and attempt to resolve, questions of *relative* human value. Thus Menenius judges Coriolanus in relation to the tribunes as one "who, in a cheap estimation, is worth all your predecessors" (*Coriolanus*, 2.1.90–91). Warwick in *2 Henry IV* suggests optimistically that Hal's reveling followers will serve "as a pattern or a measure . . . / By which his Grace must mete the lives of other" (4.4.75–76). Benvolio urges Romeo to judge Rosaline against other women: "[In] scales let there be weighed / Your lady's love against some other maid" (*Romeo and Juliet*, 1.2. 96–97). Troilus, betrayed by Cressida, fights the temptation to measure all women by her example or, as he puts it, to "square the general sex / By Cressid's rule" (*Troilus and Cressida*, 5.2.132–33).

The stakes in making such assessments seem to be highest, however, when Shakespeare's characters attempt to gauge *themselves* in relation to others, or vice versa. Celia accuses Rosalind of an "inequality" in their affections: "[T]hou lov'st me not with the full weight that I love thee" (*As You Like It*, 1.2.8–9). Falstaff accuses old men of judging the "capacities of

---

[10] Throughout this book, I sometimes refer to "Man" when I allude to classical or Renaissance concepts of humankind, which can include both men and women, although male perspectives implicitly set the standard for all. My usage is an effort to reproduce earlier terms of thought and is in no way intended thoughtlessly to subsume women within or erase them from an older paradigm of what is human. For a preliminary discussion of the way Shakespeare may have measured gender relations, see chapter 3, n. 25.

us that are young" by their own, emotional standards: "You do measure the heat of our livers with the bitterness of your galls" (2 *Henry IV*, 1.2.174–76). In *Much Ado about Nothing* Leonato declares an affinity only to another father who has so loved his child: "Measure his woe the length and breadth of mine" (5.1.11). Shakespeare reveals how often men and women make themselves the measure of others' experiences, as when the Widow in *Taming of the Shrew* suggests, telling Katherine, "Your husband, being troubled with a shrew, / Measures my husband's sorrow by his woe" (5.2.28–29); when Benvolio gauges his friend's melancholy by "measuring [Romeo's] affections by [his] own" (*Romeo and Juliet*, 1.1.126); when Timon declares, "I weigh my friend's affection with mine own" (*Timon of Athens* 1.2.216); or when Beatrice appraises Benedict, "I measure him . . . by my own spirit" (*Ado*, 2.3.143). What is striking about Shakespeare's measurers is how often they discover, by these methods, nothing so much as their own devices. As the poet of the *Sonnets* avers, "[T]hey that level / At my abuses reckon up their own / I may be straight though they themselves be bevel" (sonnet 121.9–11).

This is precisely the problem of the *homo mensura*, Man (or rather, men and women) *as* measure. As the early seventeenth-century antiquary John Selden is reported as saying about the "Measure of Things," "*We measure from ourselves*; and as things are for our use and purpose, so we approve them" (emphasis added). This is true, Selden suggests, when we attempt to compare "things" such as pears or medlars (see the epigraph to this book) but also when we compare people, as Selden illustrates with a humorous riff on Thomas Nashe: "We measure the Excellency of other Men, by some Excellency we conceive to be in ourselves. *Nash* a Poet, poor enough (as Poets us'd to be), seeing an Alderman with his Gold Chain, upon his great Horse, by way of scorn, said to one of his Companions, do you see yon fellow, how goodly, how big he looks; why, that fellow cannot make a blank Verse." It is true even of our efforts to fathom what is beyond our apprehension, "Nay we measure the goodness of God from ourselves, we measure his Goodness, his Justice, his Wisdom, by something we call Just, Good, or Wise in ourselves."[11] Shakespeare shares Selden's skepticism toward our tendency to "measure from ourselves," with the result that we keep "reckoning up" our own merits and demerits, rather, perhaps, than those of others. In terms long esta-

---

[11] John Selden, *The Table-Talk of John Selden* (London: William Pickering, 1847), 121–22.

blished by the history of epistemological skepticism, from the ancients all the way to contemporary philosophies of science, Shakespeare, in his concern with measurement, focuses on the problem of the "criterion," the difficulty of legitimating or certifying the criteria, standards, or norms by which judgments are made. (See chapter 1 for a discussion of the philosophical and material contexts for understanding sixteenth-century skepticism concerning the "criterion," or the problem of "measuring" the measure.) For Shakespeare, the chief criterion in question is Man himself, the *homo mensura* who makes himself the measure, standard, or norm by which others must be appraised.

Such a man, as Shakespeare represents him, seeks to know whether others are "more," "less," or "equal" to him by selecting the basis, and the means, by which they are to be compared. Throughout this book, I return to what may be the most urgent of these human measurements—the uses and abuses of the notion of human "equality." My chapters examine a range of early modern criteria for determining equality, including standards set by mathematics, theology, medicine, and law, as applied by those in a position to "rule." These investigations elucidate the Renaissance meanings of key phrases in the history of the term, including "equality by nature" (chapters 3 and 4) and "equality under the law" (chapter 5). Whatever Shakespeare's own ideals for human relations might have been, he contends that equality, by Renaissance standards, is no more "self-evident" than any other result of a comparative human measurement. Rather, Shakespeare reveals the extent to which all early modern determinations of human worth, of who we are and what we deserve, in relation to others, depend on which norms, and whose, define the comparison. For Shakespeare, Renaissance "equality" is just another, contingent, measure of man.

I would like to conclude by considering Shakespeare's best-known and most explicit discussion of human value, the Trojan debate over Helen's "worth" in act 2, scene 2, of *Troilus and Cressida.* Or rather, by reconsidering it: the scene has often been profitably discussed in the context of an emergent, early modern capitalism in which the contingencies of Helen's value depend on new terms of commodification, commerce, and exchange.[12] Important as the discourse of Renaissance

---

[12] In chapter 4, "Shakespeare's Social Arithmetics," I consider Shakespeare's use of Renaissance economic measures, especially the sources of his ideas about economic "equality." For recent work on Shakespeare's engagement with Renaissance economics, see Linda Woodbridge, ed., *Money and the Age of Shakespeare: Essays in New Economic*

economics is to the play, Shakespeare sets his questions about human value in a much older philosophical context, as Hector indicates when he suggests that Aristotle would have deemed Troilus, in his arguments about Helen, "unfit" for "moral philosophy" (167). With the war having reached a bloody impasse, the sons of Priam debate the merits of returning Helen to the Greeks and to her husband, Menelaus. But their particular, political deliberations give way to a dialogue that tends toward the abstract and the academic, as Troilus asks, philosophically, "What's aught but as 'tis valued?" (52). The Trojans, despite their differences, agree: nothing is, and moreover no *one* is, except insofar as their value is determined.

The quarrel begins when Hector challenges the Trojans' presupposition that Helen, however lovely, however rare, equals or exceeds the many who have lost their lives to defend her:

> Since the first sword was drawn about this question,
> Every tithe soul, 'mongst many thousand dismes,
> Hath been as dear as Helen; I mean, of ours.
> If we have lost so many tenths of ours,
> To guard a thing not ours nor worth to us
> (Had it our name) the value of one ten,
> What merit's in that reason which denies
> The yielding of her up?
> (2.2.18–24)

Hector sets up the problem of Helen's value in terms of relative worth. Designating hers as lower than "the value of one ten," he calculates that she is worth far less than the "numbers" of those who have died for her ("disme" comes from early English mathematics and refers to "decimal" arithmetic, by tenths).[13] For Hector, there is no "merit"—again, "worth" or "value"—in their "reasons." The word "reason," in early modern English, often denoted mathematical reasoning or calculation (see chapter 4).

Criticism (New York: Palgrave Macmillan, 2003). Joel Kaye's *Economy and Nature in the Fourteenth Century* remains the best groundwork for understanding early modern notions of economic value and exchange. For the rhetoric of economics in Renaissance literature, see Sandra K. Fischer, *Econolingua: A Glossary of Coins and Economic Language in Renaissance Drama* (Newark: University of Delaware Press, 1985).
[13] *Oxford English Dictionary*, s.v. "disme."

In response, Troilus accuses Hector of acts of "mismeasure." His re-
proval ranges across the terms of number, weight, and linear measure:

> Weigh you the worth and honor of a king
> So great as our dread father's in a scale
> Of common ounces? Will you with compters sum
> The past-proportion of his infinite,
> And buckle in a waist most fathomless
> With spans and inches so diminutive?
> (26–31)

Priam is "fathomless," according to Troilus, too "great" to be measured
in "common ounces," common numbers, "spans," or "inches." Troilus's
doubts regarding human measurements center, especially, on the ancient problem of the criterion. He asks: How can any standard prove sufficient in the judgment of (great) men and women, who are incompara-
ble? Helen, too, he insists, is "inestimable" (88).

Nonetheless, Troilus and Hector ultimately agree on a criterion for
continuing the war, a "reason" for the deaths of their countrymen. Hector argues that value inheres in the thing or the person being measured
and not in the measurer:

> [V]alue dwells not in particular will,
> It holds his estimate and dignity
> As well wherein 'tis precious of itself
> As in the prizer.
> (53–56)

Observing that Hector had once deemed Helen "worthy" enough to
fight a war over, Troilus attempts to exploit Hector's theory of intrinsic
value to argue that his brother cannot just simply change his mind on
the matter, that Helen's original price must be fixed:

> If you'll confess [Paris] brought home worthy prize—
> As you must needs, for you all clapp'd your hands,
> And cried, "Inestimable!"—why do you now
> The issue of your proper wisdoms rate,
> And . . .

Beggar the estimation which you priz'd
Richer than sea and land?
(86–92)

Hector stands firm, however, in his charge that Troilus (and Paris) are so
bound to their particular wills that there is no way for them to make what
he calls, in a significant turn of phrase, a "free determination" (170) in
the question of Helen's value.

But then, even as Hector maintains his right, he comes to a sudden
rapprochement with his brothers:

Hector's opinion
Is this in way of truth; yet ne'er the less
. . . I propend to you
In resolution to keep Helen still,
For 'tis a cause that hath no mean dependance
Upon our joint and several dignities.
(188–93)

Troilus and Hector finally agree to a standard that sanctions a common
resolution to the problem: their *own* honor or "dignity" (from Latin *dig-
nitat-em*, "merit," "worth"; hence, social rank or estimation). Troilus en-
dorses this ultimate "determination" wholeheartedly: "Why, there you
touch'd the life of our design!" (194). Hector never backs down from the
"truth" that Helen's value is less than what it once was, or rather seemed
to be, yet he will defend her original estimation for the sake of theirs. In
other words, *their* value "hath no mean dependance" on it. These men are
the measure of her worth, and she in turn becomes the measure of theirs,
in an infinite, circular, reflexive application of criteria that defers the discovery of "true" value, if indeed such a discovery is ever possible. (Shakespeare implies that they may have been right, in the first place, to assess
Helen as "inestimable," in the sense that it is impossible accurately to
measure her, or Priam, or anyone else for that matter.) In *Troilus and
Cressida*, Shakespeare suggests that there are no "free determinations" in
the measures of men. But even more disturbingly, Shakespeare reveals
how men may choose to apply criteria they know to be arbitrary or even
false, if it's to their own advantage to do so. And the consequences in this
case, as Hector knows from the start, include the loss of thousands of
human lives now held to be worth, in relation to Helen's, nothing at all.

Protagoras, according to Plato, espoused a "secret doctrine" whereby "Nothing is, on its own, any one thing"; for example, he claimed that nothing can be "heavy" except in relation to something else.[14] In his experiments with human measurements, Shakespeare seems to share this secret. The idea that human value is determined by potentially shifting human "relationships" is sometimes rendered comically, sometimes tragically, in *Troilus and Cressida*. The play begins with Pandarus's farcical and unabashed efforts to raise Troilus's worth, in Cressida's eyes, by comparing him favorably to all of the heroes on parade: "Troilus is the better man of the two" (1.2.60–61); "Hector is not a better man than Troilus" (79–80); "Helen loves him [Troilus] better than Paris" (107–8); "[M]ark Troilus above the rest" (183–84). *Troilus* betrays, again and again, how far and how precipitously a man or woman's worth can change as a result of new comparisons. Achilleus, for example, suffers a radical depreciation over the course of the play when his "[i]magin'd worth" (2.3.172) is deflated by Ulysseus' calculated effort to raise that of Ajax.[15] (Ulysseus' manipulations cast a cynical light on his famous speech on "degree," in which he purports to claim nature as the basis of men's ranked—and *fixed*—relationships to others. He, of all people, knows how easy it is to manipulate degree and how expedient it can be to do so.) *Troilus and Cressida* offers an example of how Shakespeare's measures make human identity dependent on human *relations*. However flawed he found the terms it set for judging people, Shakespeare, in his rhetoric of measurement, disclaims the idea of discrete human natures and proceeds by correlation, as if to say that all we can know of ourselves is by reference to others, and all they can know by respect to ourselves.[16]

---

[14] See Plato, *Theaetetus*, 122n1b.

[15] See C. C. Barfoot, "*Troilus and Cressida*: 'Praise us as we are tasted,'" *Shakespeare Quarterly* 39.1 (Spring 1988): 45–57, for a discussion of the relationship between the words "praise," "prize," and "price" in the play.

[16] What distinguishes Shakespeare's ideas about measurement as early modern rather than classical is his sense of the radical contingencies of value. As Joel Kaye has explained, the quantifying spirit that gave rise to new methodologies in the arts and sciences changed an earlier worldview "from a static world of numbered points and perfections to a dynamic world of ever-changing values conceived as continua in expansion and contraction; from a mathematics of arithmetical addition to a mathematics of geometrical multiplication, newly accepting of the approximate and the probable; from a world of fixed and absolute values to a shifting, relational world in which values were understood to be determined relative to changing perspectives and conditions; and from a philosophy focused on essences and perfections to one dominated by questions of quantification and measurement in respect to motion and

In my first chapter, I will provide sixteenth-century touchstones for understanding Shakespeare's perspectives on measurement, before proceeding to chapters exploring his specific applications of Renaissance poetic, theological, medical, mathematical, legal, and intellectual measures to human concerns.[17] All together, the chapters in this book reflect on why, given, especially, the positivist outlook of early, quantifying sciences, Shakespeare's measures are so often represented as "mismeasures," why his rhetoric of measurement, rather than create certainty, so often signifies doubt. This book is about early modern criteria for judging and comparing people, Renaissance "measures of Man," and why, for Shakespeare, they remain as equivocal, provisional, and unreliable as the men and women they are designed to assess.

---

change" (Kaye, *Economy and Nature*, 1). Shakespeare portrays a world of human values that are analogously "dynamic," "ever-changing," "shifting," and above all, "relational"—dependent on "changing perspectives and conditions," including what or whom people are judged against. For a discussion of the persistence of Aristotelian ideas about value in the early modern world, see Joshua Scodel, *Excess and the Mean in Early Modern English Literature* (Princeton: Princeton University Press, 2002).

[17] Although I have tried to stretch the reach of this book across a variety of Renaissance measures, drawn from a number of Renaissance disciplines, it does not offer a comprehensive "metrology" of the period. Left unaccounted for here, for example, are other economic measures, including money itself, land and cartographic measures, and the measure of time; these too may have conditioned Renaissance conceptions of human value. My principle of selection was rhetorical: I have focused here on Shakespeare's texts that have the greatest concentration of measuring terms, that is, where a discourse of measurement clearly informs the larger structure of his work. Fragmentary as my study is as a survey of early modern measuring systems, I hope it makes a coherent case for Shakespeare's place in the Renaissance "culture of measurement."

THE RENAISSANCE ART
OF MEASUREMENT

THE MOST FAMOUS REPRESENTATION of a perfect parity between Renaissance art and Renaissance science is no doubt Leonardo da Vinci's "Vitruvian Man," an illustration of the ancient architect's mathematical specifications for the proportions among the parts of the human body. Leonardo took careful account, in his notebooks, of Vitruvius's human measurements; indeed, the drawing may be said to embody Leonardo's own version of Protagoras's *homo mensura,* a man who at once measures and is measured by the magnitude of the world. Nature herself, Leonardo writes, framed the ratios that formulate human anatomy:

> [T]he measurements of the human body are distributed by Nature as follows: that is that 4 fingers make 1 palm, and 4 palms make 1 foot, 6 palms make 1 cubit; 4 cubits make a man's height. . . . The length of a man's outspread arms is equal to his height. From the roots of the hair to the bottom of the chin is the tenth of a man's height; from the bottom of the chin to the top of his head is one eighth of his height; from the top of the breast to the top of his head will be one sixthe of a man. From the top of the breast to the roots of the hair will be the seventh part of the whole man. From the nipples to the top of the head will be the fourth part of a man . . . (etc.).[1]

Whatever Leonardo's own understanding of the relationship between art and mathematics, with hindsight we know there's a problem: in taking

---

[1] Leonardo da Vinci, *The Notebooks of Leonardo da Vinci,* 2 vols., ed. Jean Paul Richter (New York: Dover Publications, 1883), 1:182, entry 343.

the measure of Man, even by the most scientific methods available, Leonardo was not revealing a truth so much as picturing it—creating an image, promoting a fiction, of what we might call a "human geometry." Shakespeare, among his contemporaries, was more deliberate in his efforts to show that the *homo mensura,* when configuring people, was an artist at best, a prevaricator at worst; either way, the man who measures other men often imagines or projects what he claims to know about them.[2]

Historians of science have chronicled an epistemological revolution in early modern thought, a transformation in both the methods and the aims of scientific inquiry toward "the measure of all things." Peter Dear, for example, relates the rise of "[a] mathematical philosophy that had ambitions to the measurement of all things," "a quantitative epistemology, which held that such an ideal exhausted everything accessible to human knowing."[3] Renaissance natural philosophers, the forerunners of our modern scientists, often operated on the theory that "nothing is in principle unmeasurable."[4] As I have suggested in the introduction, the impressive number of measuring words in his poems and plays argues Shakespeare's participation in this revolution, his terms of engagement with principles of Renaissance pantometry or "measuring all."

It is essential to note, however, that many Renaissance thinkers raised or revived challenges to the quantifications, standardizations, or "normalizations" achieved (or attempted) via early modern measurements. Rather than reviewing the significance of methodologies of measurement in the development of modern science, a topic well rehearsed in recent scholarship, I will focus in this chapter on something that is less familiar to us—the coincident, contemporary sense of the difficulties and even the dangers of creating, and applying, the measures of men. De-

[2] Many of Shakespeare's characters may be understood as artists who measure from themselves and create their own "truths" in the process. As later chapters in this volume bear out, the trouble comes when such characters fail (deliberately or not) to acknowledge the imaginary and provisional nature of their measures and posit their fancies as inalienable and unalterable "facts." See Elizabeth Spiller, *Science, Reading, and Renaissance Literature* (Cambridge: Cambridge University Press, 2004) for an exploration of early scientists as "makers" of knowledge.

[3] Peter Dear, *Discipline and Experience: The Mathematical Way in the Scientific Revolution* (Chicago: University of Chicago Press, 1995), 9, 1.

[4] Rudolf Carnap, *An Introduction to the Philosophy of Science,* ed. Martin Gardner (New York: Basic Books, 1966), 105.

spite the claims of early modern scientists to inductive discovery, Shakespeare had reason to imagine that "measurement" was always and already a matter of art rather than empirical science, a practice of making the rules that then prescribe, rather than reveal, the dimensions of human knowledge. (Shakespeare seems to have been ambivalent toward the "creative" nature of human measurement, as later chapters in the book attest.) In this chapter, I consider some of the trouble with Renaissance measures, as Shakespeare's contemporaries experienced it. Along with their renewed interest in a skeptical tradition that focused, precisely, on faulty measures as a way of representing the limits of human apprehension, Renaissance men and women also had material reasons to distrust them. Providing some key points of reference for a new Renaissance "metrology"—an old and enduring field of inquiry traditionally confined within the disciplines of historical science—this chapter calls for a reassessment of the early modern "culture of measurement."

Protagoras's *homo mensura* has long been identified as a progenitor of Renaissance humanism, with its emphasis on Man and his capacities. His contribution to the ethos of humanism often takes two, opposing, forms: on the one hand, the homo mensura is expressive of an early modern philosophical predisposition to celebrate the scope of human cognition; on the other, a contrary predisposition to concede or lament its bounds. If Protagoras's man-measure spurs Dr. Faustus's desire for a dominion that stretches "as farre as doth the mind of man,"[5] he also enables Hamlet's determination of Denmark as a "prison," "too narrow" a world insofar as his own mind measures it, for "[t]here is nothing either good or bad, but thinking makes it so" (2.2.249–50). Ambivalence regarding the powers and the limitations of the human mind haunts the reception of the homo mensura, from Socrates through Francis Bacon, the early seventeenth-century philosopher and essayist often held to be the father of modern empirical science. In the first part of this chapter, I will trace the transmission of the homo mensura from ancient to Renaissance epistemologies, with an emphasis on Montaigne—perhaps the most direct link to Shakespeare's own judgments about human meas-

---

[5] Christopher Marlowe, *Dr. Faustus*, in *The Complete Works of Christopher Marlowe*, 2 vols., ed. Fredson Bowers (Cambridge: Cambridge University Press, 1973), vol. 2, 1.1.87–88.

urements. Although this is a narrative that may belong, principally, to a larger history of early modern European skepticism,[6] my attention here is necessarily and purposefully limited to Protagoras's rhetorical legacy, that is, to those authors and texts that explicitly cite the homo mensura and consider the consequences of identifying man and his capacities with "measures."

There is no way of knowing, of course, what exactly Protagoras meant when he talked about man as the "measure of all things." In the *Theaetetus*, Plato has Socrates paraphrase Protagoras this way:

> [H]e [Protagoras] says, you remember, that a man is the measure of all things. . . . And he means something on these lines: everything is, for me, the way it appears to me, and is, for you, the way it appears to you; and you and I are, each of us, a man.[7]

As a prelude to his theory of Forms, Socrates refutes Protagoras by challenging the truth of appearances that vary from man to man. As Plato put it in the *Laws*, it is not man but "God [who is] 'the measure of all things' in the highest degree—a degree much higher than is any 'man.' "[8] Aristotle, in the *Metaphysics*, echoes Plato in his assessment of Protagoras:

> [H]e [Protagoras] said that man is the measure of all things, by which he meant simply that each individual's impressions are positively true. But if this is so, it follows that the same thing is and is not, and is bad and good, and that all the other implications of opposite statements are true . . . if reality is of this nature, everyone will be right.[9]

---

[6] For a comprehensive study of skepticism in the Renaissance, see Richard H. Popkin, *The History of Scepticism from Erasmus to Spinoza* (Berkeley: University of California Press, 1979). See also Victoria Kahn, *Rhetoric, Prudence, and Skepticism in the Renaissance* (Ithaca: Cornell University Press, 1985), esp. chapter 5, "Montaigne: A Rhetoric of Skepticism," which includes an excellent discussion of Montaigne's preoccupation with "scales" as a figure for judgment (115–51).

[7] Plato, *Theaetetus*, trans. John McDowell (Oxford: Clarendon Press, 1973), 152a.

[8] Plato, *Laws*, trans. R. G. Bury, Loeb Classical Library, 2 vols. (Cambridge, MA: Harvard University Press, 1926), vol. 1, bk. 4, 716C.

[9] Aristotle, *Metaphysics*, trans. Hugh Tredennik, Loeb Classical Library (London: William Heinemann, 1933), XI.vi.1–2.

Aristotle dismisses the utility of Protagoras's pronouncement altogether:

> Protagoras says that "man is the measure of all things," meaning, as it were, the scholar or the man of perception; and these because they possess, the one knowledge, and the other perception, which we hold to be the measures of objects. Thus, while appearing to say something exceptional, he is really saying nothing.

The homo mensura has been overrated, according to Aristotle; Protagoras said nothing more profound than that men perceive the world around them. Insofar as he distinguished the agent from the object of intellectual measurements, Aristotle contends, Protagoras actually had it backward; it is not we who measure the world but the world that measures us (in Aristotle's interpretation, "measure" means "causes to know"): "We also speak of knowledge or sense-perception as a measure of things . . . because through them we come to know something; whereas really they are measured themselves rather than measure other things."[10] While he critiques Protagoras's theory of human understanding, Aristotle preserves his terms, especially his use of "measurement," to articulate the ways we attain knowledge.

The rhetoric of measurement figures importantly in medieval epistemologies, as evidenced by Thomas Aquinas's treatise *Truth (De veritate)*. Aquinas returns directly to Aristotle's critique of Protagoras, asking, "Is Truth Found principally in the Intellect or in Things?" He argues that the human mind is capable of taking accurate measures, but only of the mind's own works, its own creations. As far as God's works are concerned, Aquinas follows the Philosopher:

> It is clear, therefore, that, as is said in the *Metaphysics*, natural things from which our intellect gets its scientific knowledge measure our intellect. Yet these things are themselves measured by the divine intellect. . . . The divine intellect, therefore, measures and is not measured; a natural thing both measures and is measured; but our intellect is measured, and measures only artifacts, not natural things.[11]

---

[10] Ibid., IV.v.1, XI.vi.1–2.
[11] Thomas Aquinas, *Truth*, 3 vols., trans. Robert W. Mulligan (Indianapolis: Hackett Publishing, 1954), vol. 1, question 1, article 2, p. 11.

Like his classical predecessors, Aquinas privileges God's measures over men's, but his case rests on the Bible (and its Apocrypha), especially the doctrine of Solomon's Wisdom: "Thou hast ordered all things in measure, and number & weight." Aquinas was among the many Christian critics of Protagoras who would find, in God, the one true criterion, the sole guarantor of truth.

It is the fifteenth-century philosopher Nicholas de Cusa, however, who offers the most sustained analysis of Protagoras's homo mensura, what it means for man to measure. Although he concurs with Aquinas that God is the ultimate measure, he insists that the mind of man is a creditable "measure" in His image. In his four books that portray an *Idiota* (someone without knowledge of Latin, or, perhaps, a "layman") teaching a philosopher about knowledge, Cusanus identifies the human mind with measuring in the most literal sense, as he traces its linguistic derivation. "Mind is that from which comes the limit and measure of all things. In fact, I propose that 'mind' is so called from 'measuring' [mentem quidem a mensurare dici coniicio]."[12] The etymological link between *mens* and *mensurare* reveals exactly what it is the mind performs:

> Mind is named from measure so that calculating measurements is the basis for the name. . . . Mind constructs the point as the limit of line and line as the limit of surface and surface as the limit of body. It constructs number and thus multitude and magnitude stem from mind. Hence it measures everything.

The mind, as Cusanus imagines it, is not a fixed measure but one that adapts itself to whatever it seeks to gauge:

> Notice that the mind is a kind of absolute measure which cannot be greater or smaller since it is not restricted to quantity. When you note the mind is a living measure that measures by itself (as if a living pair of compasses were to measure by themselves), then you grasp how it fashions a concept, measure, or exemplar.[13]

---

[12] Nicholas de Cusa, *Idiota de Mente*, trans. Clyde Lee Miller (New York: Abaris Books, 1979), 43.
[13] Ibid., 71, 75.

For Cusanus, the human mind, as a flexible, "living measure," "fashions" itself according to what it seeks to know.

Cusanus also wrote extensively about "real" measuring instruments, such as the balance. In his *De staticis experimentis* (Of experiments with weight), he elaborates on the idea that measuring is the basis of all our knowledge about the created world: "Although nothing in this world can be read with precision, yet wee finde by experience, the judgement of the Balance, one of the truest things amongst us. . . . [B]y the difference of weights, I thinke wee may more truly come to the secret of things." The subject of *De beryllo* (Of the lens), in turn, is "the lens of man's mind which by measuring discovers the unseen truth of the divine glory." Citing Protagoras's homo mensura directly, Cusanus describes how "man finds in himself as if in a measuring reason everything that has been created."[14] Once again, Cusanus aims to prove how human measures imitate, and strive toward, divine ones. When we take the weight of things, for example, we emulate the God who, according to Proverbs 8.28, "weighed the fountains of waters, and the greatnesse of the Earth, as in a Balance." Yet when the philosopher asks the *Idiota* why the mind measures, he says nothing of God: "It does so to reach its own measure. Mind is a living measure which achieves its own capacity by measuring other things. Everything it does is to come to know itself."[15] The recursivity of the *mens-mensura*, for Cusanus, does not subvert the mind's claims to "real" knowledge; rather, it marks *self*-knowledge as our decisive achievement.

His contemporary, the painter, architect, sculptor, musician, and philosopher Leon Battista Alberti, concurred that the mind that measures discovers, above all, its own latitude, its own span. In Protagoras's homo mensura, Alberti saw an ancient expression of a humanist confidence in the dignity of Man. His *Libri della famiglia* (ca. 1430) glosses Chrysippus's belief that everything was made to serve man by observing that "Protagoras, another ancient philosopher, seems to some inter-

---

[14] Quoted in Charles Trinkhaus, "Protagoras in the Renaissance: An Exploration," in *Philosophy and Humanism: Renaissance Essays in Honor of Paul Oskar Kristeller*, ed. Edward P. Mahoney (New York: Columbia University Press, 1976), 201–2. I am indebted to this article for its account of the reception of Protagoras in the Renaissance. I have expanded Trinkhaus's citations from the period and conclude that knowledge of Protagoras was far more widespread in England than he indicates in his survey.

[15] Cusa, *Idiota de Mente*, 75.

preters to have said essentially the same when he declared that man is the mean and measure [modo e misura] of all things."[16] In his *De pictura* (1435), Alberti suggests that Protagoras established man as a standard against which everything else may be apprehended, a criterion for comparison: "Comparison is made with things most immediately known. As man is the best known of all things to man, perhaps Protagoras, in saying that man is the scale and measure of all things, means that accidents in all things are duly compared to and known by the accidents of man." Alberti compared the homo mensura to the painter, whose perspective provides its own "measure" of reality, as art historians have explained: "[according to Alberti] the human figure is not just a convenient measuring rod for the painter to get his dimensions into accurate scale, but the space will be subjectively viewed and measured by the human observer." For Alberti, the art of painting not only depends on measuring the human form but on the measurements of the observer himself, who sees his model in relation to his own physical and psychological dimensions.[17] The reflexive nature of human measurement, once again, means that the artist, in seeking to gauge another, assays himself. For Alberti, however, as for Cusanus, this implies no failure of apprehension or limitation to it; on the contrary, it dignifies the artist's perspective in its creative range and reach.

The Renaissance gained a wider knowledge of Protagoras's philosophy through Marsilio Ficino's translations of Plato's *Protagoras* and *Theaetetus* (1466–68) and the rediscovery of works by the third-century AD skeptic philosopher Sextus Empiricus.[18] Ficino rehearsed Plato's original objections to Protagoras, including his assertion that God, not man, is the measure: "By these words Plato seems to confute Protagoras' saying that man is the measure of things."[19] No doubt, however, it was Sextus's treatise "Against the Logicians"—a skeptical refutation of all those who professed any kind of technical knowledge—that provided the most detailed account and assessment of the homo mensura available in

---

[16] Leon Battista Alberti, *Libri della famiglia*, trans. Renée Neu Watkins (Columbia: University of South Carolina Press, 1969), 133.

[17] See Trinkhaus, "Protagoras in the Renaissance," 196–97.

[18] The works of Sextus Empiricus were largely unknown before 1562, when Henry Estienne published the first Latin edition of the *Hypotyposes*. Further works were soon published by Gentian Hervet and others. For a full account of the transmission of Sextus, see Popkin, *History of Scepticism*.

[19] Ibid., 206.

the sixteenth century. Sextus based his skepticism on the status of what he called the three "criteria" of knowing:

> For just as in the process of examining heavy and light objects there are three criteria, the man who weighs, the scales, and the act of weighing, and of these the weigher is the criterion of the agent, the scales that of the instrument, and the act of weighing that of the use; and again, just as for the determination of things straight and crooked there is need of a craftsman and a rule and the application of the rule; so, in the same way, in philosophy also, for the determination of things true and false, we require the three criteria.[20]

"Measures" may include real instruments as well as cognitive ones: "Every measure or standard of apprehension . . . includes every technical measure of apprehension, so that in this sense one would call the cubit, the balance, the rule and the compass 'criteria.' " Sextus, however, maintains that our internal "criteria" are worthy of special attention in any epistemological inquiry: "It would indeed be monstrous if, while spending the utmost pains in investigating the external criteria—such as rules and compasses, weights and scales—we should neglect the Criterion within us." Protagoras should be counted "among the company of those philosophers who abolish the criterion, since he asserts that all sense impressions and opinions are true and that truth is a relative thing inasmuch as everything that has appeared to someone or been opined by someone is at once real in relation to him." Yet it is impossible, Sextus admits, positively to refute the idea that man is the measure of all things, "[f]or if anyone shall assert that man is not the criterion of things he will be confirming the statement that man is the criterion of all things; since the very person who makes the assertion is himself a man, and in affirming what appears relatively to himself he confesses that this very assertion of his is one of the appearances relative to himself."[21]

Sextus's skepticism challenges the validity of all three "criteria"—the measurer, the measures themselves, and acts of measurement. The problem, in each case, is that measures must *themselves* be measured if they

[20] Sextus Empiricus, *Against the Logicians*, trans. R. G. Bury, Loeb Classical Library (London: William Heinemann, 1976), I.36.
[21] Ibid., I.31–32, 27–28, 60–61, 340–41.

are to be held accountable as standards of truth. Measurement necessitates an infinite regress of operations:

> Those who claim for themselves to judge the truth are bound to possess a criterion of truth. This criterion, then, either is without a judge's approval or has been approved. But if it is without approval, whence comes it that it is trustworthy? . . . And if it has been approved, that which approves it, in turn, either has been approved or has not been approved, and so on *ad infinitum*. . . . [T]he proof which is adduced to confirm the criterion must needs be supported by a criterion; so that we are plunged into circular reasoning, the criterion on the one hand awaiting confirmation by the proof, and, on the other hand, the proof waiting for the support of the criterion.

Taking Protagoras's dictum to its logical conclusion leads Sextus to deduce that man is no measure or at least that "no one is definitively the criterion of truth."[22]

Protagoras's "Man is the measure" may have been axiomatic in Renaissance England, as John Selden's remarks on the phrase, for example, suggest.[23] Michel de Montaigne's discussion of the limits of human knowledge in his *Essays* (1575), however, probably represents the most immediate source of Shakespeare's own discourse of measurement. In his essay titled "It is folly to measure the true and false by our own capacity," Montaigne writes, "There is no more notable folly in the world than to reduce these things to the measure of our capacity and competence."[24] His "Apology for Raymond Sebond," which uses skepticism concerning human knowledge to argue our dependence on God, cites Protagoras directly: "Protagoras thought that 'what seemed to each man was true for each man.'" Protagoras was wrong, he explains,

> Since our condition accommodates things to itself and transforms them according to itself, we no longer know what things are in

---

[22] Ibid., I.336.
[23] John Selden, *The Table-Talk of John Selden* (London: William Pickering, 1847), 121–22.
[24] Michel de Montaigne, "It Is Folly to Measure the True and False by Our Own Capacity," in *The Complete Essays of Montaigne*, trans. Donald M. Frame (Stanford: Stanford University Press, 1958), 132.

truth; for nothing comes to us except falsified and altered by our senses. When the compass, the square, and the ruler are all off, all the proportions drawn from them, all the buildings erected by their measure, are also necessarily imperfect and defective.

> 'As in a building, if the rule is false,
> If a bent square gives verticals untrue,
> And if the level is all askew,
> Then all must be defective and at fault'
> —Lucretius[25]

Skepticism toward the homo mensura was elsewhere applied, in the sixteenth century, to the question of whether the mind was sufficient to attain knowledge of God. Pierre Charron, in *Les trois veritez* (1594), for example, found Calvinism dangerous in its efforts to make men the measure of faith and its insistence that "human measuring rods" are preferable to others.[26] For Montaigne, the analogy of mind and measure was fundamental to his thinking about all human perception: "[I]f my touchstone is found to be ordinarily false, my scales uneven and incorrect, what assurance can I have in them this time more than at other times?" Our lack of self-understanding, above all, calls all criteria into question: "[H]e who understands nothing about himself, what can he understand? 'As if he could really take the measure of anything, who knows not his own' (Pliny)." Montaigne concludes, regarding Protagoras, that "[t]hat favorable proposition [i.e., that man is the measure of all things] was just a joke which led us necessarily to conclude the nullity of the compass and the compasser." He reiterates the central insight of Sextus Empiricus regarding the problem of criteria: "To judge the appearances that we receive of objects, we would need a judicatory instrument; to verify this instrument, we need a demonstration; to verify the demonstration, an instrument: there we are in a circle."[27] For all of Montaigne's emphasis on self-knowledge as the highest intellectual pursuit, as all that we can, and should, need to know, he still maintains misgivings about whether men can ever, fully and conclusively, measure themselves. And if

---

[25] Montaigne, "Apology for Raymond Sebond," in *Complete Essays*, 443, 453–54.
[26] Quoted in Popkin, *History of Scepticism*, 59.
[27] Montaigne, "Apology for Sebond," 423, 418, 454.

they cannot, the prospects for knowledge, any knowledge, is by Montaigne's reckoning nil.

What may be most striking and surprising about the Renaissance legacy of Protagoras's homo mensura is its role in the making of early modern science. Often hailed as an early exponent of modern scientific method, Francis Bacon's 1621 treatise *Novum Organum* ("New instrument," from Greek *organon*, "instrument" or "tool") proceeds rhetorically as an effort to contain or contest skepticism about human measures. Bacon cites Protagoras as among his forerunners "who despaired of finding truth . . . [and so] turn men aside to . . . a kind of ambling around things, rather than sustain them in the severe path of inquiry."[28] As for the homo mensura, "[t]he assertion that the human senses are the measure of things is false; to the contrary, all perceptions, both of sense and mind, are relative to man, not to the universe." Yet what Bacon calls "Anticipations of Nature"—our own faulty projections of what we expect to discover in the world—may give way to true "Interpretations of Nature" via the new methods of empirical science:

> Neither the bare hand nor the unaided intellect has much power; the work is done by tools as much as the hand. As the hand's tools either prompt or guide its motions, so the mind's tools either prompt or warn the intellect.

"Real" measuring instruments help us to function without the mediation of our prejudices and false notions, Bacon's famous "idols of the mind"; moreover, they rescind the case, first presented by Socrates, that the idiosyncrasies and fallibilities of *individual* minds interfere with understanding:

> Our method of discovery in the sciences is designed not to leave much to the sharpness and strength of the individual talent; it more or less equalizes talents and intellects. In drawing a straight line or a perfect circle, a good deal depends on the steadiness and practice of the hand, but little or nothing if a ruler or compass is used. Our method is exactly the same.[29]

---

[28] Francis Bacon, *The New Organon*, ed. Lisa Jardine and Michael Silverthorne (Cambridge: Cambridge University Press, 2000), aphorism 67, p. 56.
[29] Ibid., aphorism 41, p. 56; aphorism 2, p. 33; aphorism 61, p. 50.

Bacon thus engages a tradition of philosophical skepticism, centering on Protagoras, in service of his faith that the right instruments, properly applied, *can* lead to knowledge of nature.

Bacon does not question the reliability of "criteria" that take the form of compasses or levels. His skepticism, in contrast to Montaigne's, ultimately restores faith in the homo mensura, at least if he follows the right, albeit man-made, "rules." Yet Bacon does not simply replace the uncertainties of skeptical philosophy with the certitudes of modern science. Haunting his discussion, for all his faith in the advancement of learning, is the old problem: how to be sure—*really* sure—of the instruments that men create and apply. The so-called revolutions of Renaissance science, at their inception, also marked a return—to the homo mensura, his promise of intellectual progress and his potential for error. In our attempts to reconstruct a metrology of the Renaissance, we should keep in mind the extent to which early modern philosophers, natural or otherwise, questioned whether, in taking measurements, they were really moving forward or were, as Montaigne suspected, ever going in circles.

Along with absorbing the impact of a philosophical tradition that called the measures of men into question, Tudor and Stuart England faced continual problems with actual standards for physical measurement. With the aim of further expanding the bounds of traditional "metrology"—the history of weights and measures, often viewed institutionally, as national systems—I consider now the material grounds for sixteenth- and seventeenth-century skepticism over measurement, focusing, once again, on the problem of the criterion, or "measuring" the measure.

A concern about false measures, particularly in the context of trade, is at least as old as the Bible, as numerous injunctions against them suggest:

Ye shal not do vniustly in iudgement, in line, in weight, or in measure. (Lev. 19:35)

Shal I iustifie the wicked balances, and the bag of deceitful weights? You shal haue iuste balances, true weightes. (Mic. 6:11)

Thou shalt not haue in thy bagge two maner of weightes, a great & a small. Nether shalt thou haue in thine house diuerse measures, a great and a small: But thou shalt haue a right & iust weight: a perfit & a iust measure shalt thou haue, that thy dayes may be lengthened in the land, which the Lord thy God giueth thee. For all that do suche things, and all that do vnrighteously, are abominacion vnto the Lord thy God. (Deut. 25:13–16)

English law, starting with the Anglo-Saxons, had always included legislation against the misuse of measures. In the tenth century, for example, Edgar the Peaceful ordered that all measures had to conform to standards kept at London and Winchester. King Aethelred (ca. 1000) decreed that "deceitful deeds and hateful injustices shall be strictly avoided, namely untrue weights and false measures and lying testimonies."[30] King John's Magna Carta (1215) included a clause regarding the standardization of measures:

> Throughout the kingdom there shall be standard measures of wine, ale, and corn. Also there shall be a standard width of dyed cloth, russet, and haberjet; namely two ells with the selvedges. Weights are to be standardized similarly.[31]

A century later, Edward I ordered that a permanent measuring stick be made of iron to serve as a standard for the kingdom; he also ordered that the foot be standardized at one-third of a yard and an inch at 1/36 of a yard. Edward III decreed that

> none shall sell by the bushel if it be not marked by the King's seal, and that it be according to the King's Standard; and also it is ordained that he which shall be attainted for having double measure that is to say one greater to buy and another less to sell, shall be imprisoned as false, and grievously punished.[32]

Many of these laws were designed to prohibit and prosecute deliberate efforts to cheat the system(s), but early measuring practices sometimes varied because of unavoidable differences among "standards" in use. There were practical problems in creating and disseminating identical

---

[30] R. D. Connor, *The Weights and Measures of England* (London: Her Majesty's Stationery Office, 1987), xxiv.

[31] Quoted in A. E. Berriman, *Historical Metrology* (London: Dent and Sons, 1953), 560.

[32] Quoted in Connor, *Weights and Measures of England*, 327.

measures and a long history of inconsistency in their application.[33] Thus Edward I issued an edict that declared a pound to be equal to twelve ounces in the weighing of medicines, whereas a pound in all other cases was to weigh fifteen ounces. Edward III decreed punishments for the use of measures other than standard ones, "except it be in the county of Lancaster because in that country it hath always been used to have greater measures than in any other part of the realm."[34]

The inconsistencies of English measurement systems were nowhere more apparent than in the matter of weights, where centuries of confusion gave way to a crisis over standards in the sixteenth century. Corruption remained a problem, as is evident by the publication of such works as John Colyn's commonplace book, *Exposure of the Abuses of Weights and Measures for the Information of the Council* (ca. 1517). A London mercer, Colyn hoped to expose the "many falce Beames and Scales wherein hath bine found greate deceipt, as namely Beames made by insufficient workemen."[35] It didn't help that there was more than one "pound" in use during the Renaissance. Although the Tower pound, used to weigh coins, was abolished in 1527, the troy pound (used for gold, drugs, and bread) and the avoirdupois pound (as its name suggests, used for heavier objects) remained as independent measures, despite sharing the same designation, "pound."[36] Both were ultimately based on the weight of a grain of barley, the oldest English standard for judging weight equivalences. The troy pound equaled 5,760 grains and divided into twelve ounces, each weighing 480 grains. The avoirdupois pound equaled 7,000 grains, but it was divided into sixteen ounces, of 437.5 grains each. This created another anomaly: whereas the troy pound was lighter than an avoirdupois pound, a troy ounce weighed more than an avoirdupois ounce.[37] Even as late as 1638, John Penkethman, in his *Artachthos, or, a New Booke declaring the Assise or Weight of Bread*, calls attention to the il-

[33] Ibid., xxvi. As Connor describes the situation, "[Q]uite apart from deliberate fraud, the difficulty in making an accurate measure from a pattern was formidable."
[34] Hubert Hall, ed., *Select Tracts and Table Books Relating to English Weights and Measures, 1100–1742*, Camden Miscellany 15 (London: Offices of the Society, 1929), x.
[35] John Colyn, *Exposure of the Abuses of Weights and Measures for the Information of the Council* (London, ca. 1517), 51.
[36] To add to the confusion, a "merchant avoirdupois pound" was also in use at this time.
[37] J. Geoffrey Dent, "The Pound Weight and the Pound Sterling," *Folk Life* 27 (1988–89): 80–84, 81.

logic of the system: "[I]t appears, though the ounce *Troy* be heavyer than the ounce *Avoirdupois*, yet the Pound *Troy* is lighter then the pound *Avoirdupois*."[38]

Although there was no early modern effort to normalize the weight of a pound,[39] the Tudors enacted several rounds of legislation in their efforts to standardize each of its kinds. In 1497, Henry VII commissioned new standards of weight as well as volume, based on ancient standards that had survived in the Treasury, and had them distributed to forty-two cities and towns. These were almost immediately found to be faulty.[40] Elizabeth issued a new set of standards for troy and avoirdupois weights the first year of her reign, but these yet again were deemed defective. In 1574 she appointed a jury consisting of nine merchants and twelve goldsmiths to resolve the inconsistencies. On what was to prove only their first attempt at it, the jury based their standard for troy weights on those held by the Goldsmiths' company, ignoring, for example, those kept in the Exchequer. They started over again once it was clear that they needed to resolve the discrepancies among the many "authorized" standards in circulation. Elizabeth's second series of weights was issued that same year; seven years later, these were condemned as well. A new jury was impaneled whose rulings led to the production of a third series of standards in 1588. New weights, along with measures for length (such as the ell) and distance (Elizabeth also set down standards for the rod, the furlong, and the mile), were deposited in the Exchequer and distributed, once again, to promote the use of consistent standards throughout the nation.[41]

Elizabeth's numerous proclamations regarding the misuse of weights and the need to adhere to royal standards attest at once to the urgency of the contemporary problem, her determination to fix it, and the futility of many of her efforts. An early "Proclamation for Waightes" (1570) states that "the Waights commonly vsed within the Realme be vncertaine,

---

[38] John Penkethman, *Artachthos, or, a New Booke declaring the Assize or Weight of Bread* (London, 1638), sig. D2v.

[39] The troy pound remained in use until 1878, when it was declared illegal, but gold is still weighed today by troy ounces.

[40] Connor, *Weights and Measures in England*, 237–38.

[41] Ibid., 35, 239. See Witold Kula, *Measures and Men*, trans. R. Szreter (Princeton: Princeton University Press, 1986), for the broader European context of attempts to reform measurement systems in the sixteenth and seventeenth centuries.

and varying one from an other, to the great slaunder to the same, and the
deceiuing of man both Buyers and Sellers":

> Forasmuch as (by Gods Lawes and mans lawes) there hath beene
> and ought to be in all places, true, iust, and certaine Waightes and
> Measures, and the same also (by the lawes of this Realme) to be es-
> tablished, published, and made knowen to all people.[42]

Several early proclamations call for the standardization of balances
and scales as well as weights. Yet it is clear that royal decrees were not always effective, for as late as 1602 she made another "Proclamation for
Measures" which more or less rehearsed the concerns of the first. She
refers to her earlier legislation, "[F]or the auoiding of varietie and de-
ceits of Weights and Measures, diuerse Statutes, Acts and Ordinances
haue heretorfore beene made, That one Weight, and one Measure
should be vsed throughout the Realme," but notes that many nonstan-
dard ones remained in use "[which] for the most part are by long con-
tinuance of time, or for want of good keeping, or other defects of abuses
differing and not agreeing with the ancient Standard of measures." The
queen demanded accountability, ordering that the cities and towns "examine, trie, and size all other Measures of the Realme by, from time to
time as occasion shall require"; those committing acts of "nonconfor-
mity" were "upon paine to be apprehended, committed to prison, fined
and punished, as contriuers and vsers of false Measures," while the of-
fending measures themselves must be "broken & defaced."[43] Although
Elizabeth's third series survived her reign and beyond, her successors'
own "proclamations" make it clear that abuses remained common. In
1618, James I continued to press the need for surveillance, calling for
tradespeople of all kinds "[t]o appeare with all their Waights & Measures
the day and place aforesaid, to see them duely examined by our
Standerds," along with the "Beames and Ballances, that euery such per-

---

[42] Queen Elizabeth, "Proclamation for Waightes" (1570), in *A Booke Conteining All
Svch Proclamations, as were published dvring the Raigne of the late Queene Elizabeth* (London, 1618), n.p.
[43] Queen Elizabeth, "Proclamation for Measures" (1602), in *Booke Conteining All Svch
Proclamations*, n.p.

son hath or vseth."[44] The vicissitudes of "weighing" in Tudor and Stuart England make literal some of the theoretical and philosophical problems of "measuring" measures. The crisis was real.

Among material measures the most interesting, for Shakespeare, may well have been "anthropometric" ones, or linear measures based on the human body. Early modern anthropometric measures include the finger, the palm, the inch, the foot, the cubit, and the fathom; invoked recurrently by Shakespeare, these may represent the original standards of the West.[45] "Inch," for example, derives from *onyx* or *onych*, the Greek word for "nail" or "claw"; "ell" (from OE *eln*) is the length from the elbow to the tip of the middle finger; "span" (OE *span*) is the distance from the end of the thumb to the end of the little finger extended; while "fathom" (OE *faethm*) equals the distance between a man's outstretched arms.[46] Shakespeare has his playful moments with anthropometrics. When Boyet mocks Armado's declaration of love for the Princess's slipper, noting that he "[l]oves her by the foot," Dumaine extends the jest, "[h]e may not by the yard" (5.2.668–69). His obscene pun on "yard"—both a measure of three feet and an early English word for "penis"—is not the only time Shakespeare creates a comedy of Renaissance body measures; for example, Dromio of Syracuse puns on "Nell," the name of the woman who believes herself his wife: "[H]er name [and] three quarters, that's *an ell* and three quarters, will not measure her from hip to hip" (3.2.109–11, emphasis added). Anthropometrics, it would seem, makes Man the measure in the most literal and substantial sense.

For Shakespeare, as for others, however, the question would be, *which* man? Although Elizabeth standardized the English ell, for example, at 45 inches, the Scottish ell remained 37 inches and the Dutch ell 27. The problem of fixing and then applying standards of measure that were based, ultimately, on *different* bodies was considered at length by Paul Greaves in *Discourse of the Roman Foot* (1647). In his search for the "true" Roman foot he observes how many feet compete as standards:

[C]oncerning the precise quantity of this foot, there is not any one thing after which learned men have more inquired, or in

[44] King James I, "Proclamation for Waights & Measures" (1618), in *Booke Conteining All Svch Proclamations*, n.p.
[45] See Kula, *Measures and Men*, chapter 5, for a brief history of "anthropometric" measures in the West.
[46] *Oxford English Dictionary*, s.v. "inch," "ell," "span," "fathom."

which they doe lesse agree. For Budeus equals it to the Paris foot; Latinius Latinius . . . and others, deduce it from an ancient monument in the Vatican of T. Statilius. . . . several others, contend the foot on Cossutius Monument in Rome, to be the true Romane foot: . . . Lucas Paetus defines it from some brasse feet found amongst the rudera in Rome.[47]

Tellingly, Greaves cites Protagoras as the first to articulate the problem of "man-measures":

> For if that of Protagoras be true, as well in measures, as in intellectual notions, that man is [the measure of all things]: Whence Vitruvius observes, that the Latines denominated most of their measures, as their digit, palm, foot, and cubit, from the parts and members of a man: who shall bee that perfect and square man, from whom we may take a pattern of these measures, or if there be any such, how shall we know him? or how shall we be certain the Ancients ever made choice of any such? Unless, as some fancy, that the cubit of the Sanctuary, was taken from the cubit of Adam, he being created in an excellent state of perfection: So we shall imagine these digits, and palms, to have been taken from some particular man of completer lineaments then others. On the other side, if this foot may be restored by the digits, and palms of any man at pleasure, since there is such a difference in the proportions of men, that it is as difficult to finde two of the same dimensions, as two that have the same likenesse of faces, how will it be possible, out of such a diversity, to produce a certain and positive measure, consisting of an indivisibility, not as a point doth in respect of parts, but in an indivisibility of application, as all originals, and standards should do.[48]

I know of no other Renaissance text that so explicitly and thoughtfully outlines the difficulties of establishing an anatomical "norm," a "perfect and square" man, out of the diversity of human bodies. (For Shakespeare's exploration of Renaissance "body measures," see chapter 3.)

[47] Paul Greaves, *Discourse of the Roman Foot* (London, 1647), sig. B2v.
[48] Ibid., sig. B5.

There appear to have been occasional efforts in the history of British measures to acknowledge the differences among individual bodies in the development of a universal standard. David I of Scotland's Assizes of Weights and Measures (ca. 1150) declared that the "ulna" (the length of a forearm) should contain thirty-seven inches, measured by "the thowmys of iii men, that is to say a mekill man, and a man of messurable statur, and of a lytell man. The thoumys are to be mesouret at the rut of the nayll." The three thumbs were then to be averaged to find a "common" measure. A sixteenth-century scheme for finding a "standard" foot prescribed a similar method: "bid sixteen men to stop, tall ones and short ones, as they happened to come out [of a church]. . . . Their left feet one behind the other . . . gives . . . the right and lawful rood [the sixteenth part of which is] the right and lawful foot."[49] At other times, the problem of divergent bodies was resolved by making one body—that of the king—the standard for all. William of Malmesbury (1095–1143) reported that Henry I's arm was the basis for the yard, just as the French *pied du roi* allegedly derived from the foot of Charlemagne.[50] Whether or not the origins of measures in the royal person are mythic rather than historical, the idea of the monarch's body as the one "trew standard" seems conveniently designed to confirm his (or her) *natural* "rule."

As Greaves suggested, measures that come from the human body reimagine the homo mensura as a man-measure from whom knowledge of all things is given and taken directly, without the mediation of the mind. It has been suggested that "natural" measures, historically, have had some psychological and cultural staying power: "There seems to have been an overpowering desire throughout the centuries to have the measures relating to something natural rather than to something artificial."[51] English "body" measures that remain in use (notably, the foot) probably no longer connote an intimate association with the human body. Even if the material connection has been lost, however, the rhetorical link survives, a reminder of our desire for measures that have their source and their object in nature, conceived as a locus of truth. But as Shakespeare would see it, the measures of men are more likely to reveal the nature of the *particular* man, or woman, who embodies the "rules."

[49] Quoted in E. Nicholson, *Men and Measures* (London: Smith and Elder, 1912), 35, 36.
[50] Connor, *Weights and Measures of England*, 81.
[51] Ibid., 35.

I suggested earlier that Renaissance science arose within a skeptical tradition of philosophical inquiry focused on the possibilities and pitfalls of
human measurements. This tradition continues: Protagoras's legacy
lives on in the modern discipline we call the philosophy of science. Following their early modern antecedents, modern philosophers of science
have often questioned the role of measurement in the production of
knowledge; their work may be said to extend, refine, and systematize the
essential insights of Protagoras's early heirs. They may have never read
Sextus Empiricus, for example, but recent philosophers conform closely
enough to the ancient writer's scheme of criteria for scientific "knowing," which they claim depends not only on "an observable *property* of [a]
system whose 'value' will be determined" but also on "an *instrument* by
means of which the operation will be made" along with "measurement
capability (the objectivity of the observer)." Each of these criteria has
been subject to a philosophical skepticism focused, above all, on the alleged objectivity of measurement, leading some to conclude that "the
neutral attitude toward the philosophical foundation of measurement is
nothing but an illusion or self-deception."[52]

Such skepticism in measurement theory is generally associated with
"operationalism," a reorientation of scientific inquiry toward acts or operations, rather than objects, of measure. The "operational" approach to
science was first formulated by P. W. Bridgman:

We naturally ask *what* it is that we measure when we make a measurement. The answer often comes of itself—with a meter stick we
measure lengths. But what is this "length" which we thus measure?
Is it a thing independent of this particular thing that we do, so that
it would make sense to ask, "Can you really find the length by
counting the number of times you lay on a meter stick?" Is length
anything more than what we measure when we make a length
measurement according to prescription? In calling it a "what" at
all are we in the presence of anything more than a verbal ellipsis
for shortening our description of what we do?[53]

[52] Berka, *Measurement*, 16–18, 206.
[53] P. W. Bridgman, *The Way Things Are* (Cambridge: Cambridge University Press,
1959), 137.

Proponents of operationalism claim its origins in the theory of relativity, which unsettled the view of an objective, physical world, "independent of the observer whose experiments and observations were simply means of finding out how the world was constructed and by what laws its behaviour was governed. The emphasis has now shifted from the nature of the world to the operations of experiment and observation."[54] The insights of quantum theory, finally, made it inescapable: "We now know that . . . [the] instrument-of-observation and object-of-observation cannot be separated from each other."[55]

Operationalists challenge the assumption that bodies have "properties," which have "magnitudes" that exist before they are measured. In principle, they observe, anything might be designated as a property of an object and then measured:

> It would be possible to keep a standard copper ball, to let it fall from a given height on to various bodies, and to measure in each case the frequency of the note emitted on impact. This could be called, say, "the sonority" of the body, and it would be a true measurement. . . . Yet we do not measure sonorities. . . . Why? The traditionalist would say: because bodies have temperatures but they have no sonorities: but how he would know this I cannot imagine. I know of no revelation, human or divine, that declares what properties a body shall have.[56]

Contemporary philosophers of science have thus shifted the focus from measurable properties to the criteria for measuring them, including physical instruments. We need to ask, according to Bridgman, "What sort of information is an instrument capable of giving us? To what extent is what the instrument gives us colored by the instrument itself, or is the instrument capable of revealing to us something 'independent of the instrument'?" Operationalists consider, finally, the criterion of the man (or woman) behind the instruments, since "[t]he ultimate instrument is ourselves."[57] The notion that scientists themselves are the only relevant

[54] Herbert Dingle, "A Theory of Measurement," *British Journal for the Philosophy of Science* 1.1 (May 1950): 5–26, quote at 23.
[55] Bridgman, *Way Things Are*, 140.
[56] Dingle, "Theory of Measurement," 6, 11.
[57] Bridgman, *Way Things Are*, 140, 153.

criterion of measurability is known as "homocentric operationalism."[58] Such a philosophy poses the question, How can we know the measurer from the measure?

If Francis Bacon's *New Organon*, in its critique of classical and early modern methodologies of knowing, may be understood as an early manifesto of operationalism, Shakespeare, in his own investigation of the instruments of learning and of judgment, leaned toward its homocentric variety. Among the tools that men and women use in their personal assessments of the world, however, the most important for Shakespeare may well have been words, or what Bacon would later term the "Idols of the marketplace":

> Men associate through talk; and words are chosen to suit the understanding of the common people. And thus a poor and unskillful code of words incredibly obstructs the understanding. The definitions and explanations with which learned men have been accustomed to protect and in some way liberate themselves, do not restore the situation at all. Plainly words do violence to the understanding, and confuse everything, and betray men into countless empty disputes and fictions.[59]

The primacy of language among instruments of knowledge, not surprisingly, is also a fundamental tenet of modern philosophical operationalism.

As a means of breaking ourselves of the habit of assuming that physical objects actually "possess" magnitude, for example, Bridgman demanded a change in the way we *talk about* measurement itself. Rather than "talk[ing] about measuring a 'length,'" Bridgman suggests, we should talk instead about making a "length measurement. Or better still, talk about 'length measuring,' the 'length' here being an adverb."[60] His discussion of the terms of mensuration and the way they promote (mis)understanding does not proceed from science or scientific theory but derives, tellingly, from the linguistic philosophy of Ludwig Wittgen-

---

[58] Berka, *Measurement*, 208.
[59] Bacon, *The New Organon*, aphorism 43, pp. 41–42.
[60] Bridgman, *Way Things Are*, 137.

stein, who, throughout his works, uses "measurement" as an analogy for knowing.[61] Wittgenstein takes an "operational" approach to language:

> The meaning of a question is the method of answering it. Tell me *how* you are searching, and I will tell you *what* you are searching for. . . . *How* you search in one way or another expresses what you expect. Expectation prepares a yardstick for measuring the event.[62]

For Wittgenstein, language itself is a measuring device:

> Language is an instrument. Its concepts are instruments. Now perhaps one thinks that it can make no great difference which concepts we employ. As, after all, it is possible to do physics in feet and inches as well as in metres and centimetres; the difference is merely one of convenience. But even this is not true if, for instance, calculations in some system of measurement demand more time and trouble than it is possible for us to give them.[63]

The relationship of language to reality is like the application of a yardstick to an object: "The method of taking measurements, e.g. spatial measurements, is related to a particular measurement in precisely the same way as the sense of a proposition is to its truth or falsity."[64] Just as length is what is measured in measuring length, the meaning of "meaning" is referred to the ways we have of explaining it:

> What is the meaning of a word?
>
> Let us attack this question by asking, first, what is an explanation of the meaning of a word; what does the explanation of a word

[61] G. P. Baker and P. M. S. Hacker, "The Standard Metre," in *Wittgenstein: Understanding and Meaning*, ed. G. P. Baker and P. M. S. Baker (Oxford: Blackwell, 1980), 284–96, 284.

[62] Ludwig Wittgenstein, *Philosophical Remarks*, ed. Rush Rhees, trans. Raymond Hargreaves and Roger White (Oxford: Basil Blackwell, 1975), 12–14.

[63] Ludwig Wittgenstein, *Philosophical Investigations*, trans. G. E. M. Anscombe (Oxford: Basil Blackwell, 1967), 151e.

[64] Wittgenstein, *Philosophical Remarks*, 78; see discussion in Cora Diamond, "How Long Is the Standard Meter in Paris?" in *Wittgenstein in America*, ed. T. G. McCarthy and S. C. Stidd (Oxford: Oxford University Press, 2001), 104–39, quote at 109.

look like? The way this question helps us is analogous to the way the question "how do we measure a length?" helps us to understand the problem, "what is length?"[65]

And again, "What 'determining the length' means is not learned by learning what *length* and *determining* are; the meaning of the word 'length' is learnt by learning, among other things, what it is to determine length."[66] What the philosophy of measurement, as a subfield within the philosophy of science, shares with Wittgenstein's philosophy of language is an emphasis on the tautological relation between method and meaning, the determinative effects of theory on interpretation, measures on the measured. Yet, as the chapters ahead will show, Shakespeare had long since imagined the confusion and the role of language in helping to create it. In exploring Renaissance terms of measure, he was also engaging, skeptically, the "metrics" of rhetoric itself.

At the end of Shakespeare's last comedy, *Measure for Measure*, the would-be nun Isabella, nearly ruined by the nefarious Angelo, pronounces her final judgment, makes her last calculation, regarding her seducer's villainy: "Nay, it is ten times true, for truth is truth / To the end of reckoning" (5.1.45–46). It may be that Shakespeare ultimately concurred with Isabella's account of the truth as fixed and enduring. His skepticism, like Montaigne's, may have had its limits: despite the inconsistencies and insufficiencies of our efforts to determine it, his characters generally maintain a belief in a truth that transcends (at least ten times over, by Isabella's estimate) the reckonings of men. As for the human mind that measures, however, Shakespeare so often follows Montaigne: "[I]t is not easy to discover its miscalculation and irregularity . . . [It] is an instrument of lead and of wax, stretchable, pliable, and adaptable to all biases and measures; all that is needed is the ability to mold it."[67]

But it is precisely because of their "adaptability," their susceptibility to "molding," however, that Shakespeare compels us to attend to the measures of the mind, including the instrumentality of language itself. His poetics posits the "measure" of men and women as metaphorical, a figure of speech rather than an unmediated or unbiased "figuring out." In

[65] Ludwig Wittgenstein, *The Blue and Brown Books*, 2nd ed. (New York: Harper and Row, 1960), 1.

[66] Wittgenstein, *Philosophical Investigations*, 225e.

[67] Montaigne, "Apology for Sebond," 425.

the next chapter, in its specific focus on Shakespeare's literary language, I deal directly with Shakespeare's mindfulness of the "operations" of rhetoric itself, but the playwright's measuring languages are fore-grounded throughout the chapters that follow. By insisting that we re-consider the terms of measurement that help regulate our experiences, Shakespeare anticipates Wittgenstein: "Philosophy, as we use the word, is a fight against the fascination which forms of expression exert upon us."[68] In the process, Shakespeare restates some of the oldest and most confounding epistemological problems in terms of an early modern pol-itics of language, most perceptibly revealed through literature—or the art of what language makes possible.

[68] Wittgenstein, *Blue and Brown Books*, 27.

POETIC NUMBERS AND
SHAKESPEARE'S "LINES OF LIFE"

I am ill at these numbers; I have not art to reckon my groans.
—HAMLET (*Hamlet*, 2.2.120–21)

SHAKESPEARE'S CHARACTERS, ESPECIALLY when they are in love, struggle to measure their feelings, to "tell . . . how much" they love, as Cleopatra urges Antony (*Antony and Cleopatra*, 1.1.14). There may be beggary in "the love that can be reckon'd" (15), but nevertheless Renaissance English verse is frequently understood by its practitioners as the art of "numbering" one's desires. This chapter will reveal the extent to which Renaissance writers thought of poetry as a system of measurement, a means of evaluating oneself and one's world through verse, a rhetorical weighing of human lives and loves. We may be more familiar with mimetic theories of poetry, or ones that assimilate verbal art to the visual arts or to visual representation more generally (poetry as a "speaking picture," an icon, a mirror). Sixteenth-century poets, however, additionally thought of their writing as an instrument of measure, one that proceeded quantitatively rather than qualitatively, deciding relations rather than depicting "nature." Shakespeare's poetics is based on evaluations of respective worth: "Shall I compare thee to a summer's day? / Thou are *more* lovely and *more* temperate" (sonnet 18, lines 1–2, emphasis added). In this chapter, I examine two formal and aesthetic aspects of Shakespeare's poetry, the comparative "figures" of his *Sonnets* and the meters of the *Sonnets* and late plays. Both represent Renaissance attempts to measure human experience by the numbers of verse—for Shakespeare, with equivocal results.

Renaissance poetic theory regularly presents verse as an art of "enumeration," as works such as George Puttenham's *Arte of English Poesie* (1589) attest. Puttenham's treatise begins with a history of poetry and the poetic kinds and follows with two books of instruction on the making of verses. The first of these, titled *Of Proportion Poetical,* treats metrics and stanza form; the second, *Of Ornament,* concerns rhetoric and the art of verbal figuration. In each, Puttenham prescribes poetic criteria or "rules" for setting up comparisons among words, and among the people to whom they relate.

For Puttenham, "meter" is another word for poetic "numbers." His work is representative of an early modern habit of confusing *arithmos* (for Puttenham, a variant of "arithmeticall"), *rithmos,* arithmetic, and the Latin *ars metrica.* His book *Of Proportion Poetical* identifies the art of meter with the "Mathematicall sciences," those disciplines that teach that "all things stand by proportion."[1] As is well known, "proportion"—meaning "a portion or part in its relation to the whole; a comparative part; the relation existing between things or magnitudes as to size, quantity, number or the like; comparative relation"; or, in mathematical terms, "an equality of ratios"[2]—represented an ethical ideal in the Renaissance as well as an aesthetic one. Puttenham introduces his discussion of poetic proportion by citing the words of Solomon's Wisdom, "God made the world by number, measure, and weight" (64). At least one of his contemporaries identified the art of meter with the Creator's mathematics:

> The source of metre is Almighty God, inasmuch as he created this universe and everything contained within its sphere according to a fixed plan, as it were by measure. . . . All the instruments which we use are made with certain proportions, that is by measure. If this occurs in things, how much more so in language, which gives expression to all things?[3]

For Shakespeare, however, the proportions of human art do not operate according to a "fixed plan": instead of espousing a faith in "true" pro-

---

[1] George Puttenham, *The Arte of English Poesie,* ed. Gladys Doidge Willcock and Alice Walker (Cambridge: Cambridge University Press, 1970), 64. All further citations of Puttenham come from this edition and appear parenthetically in the text.

[2] *Oxford English Dictionary,* s.v. "proportion."

[3] Richard Willes, *De Re Poetica* (1573), quoted in Derek Attridge, *Well-Weigh'd Syllables: Elizabethan Verse in Classical Metres* (Cambridge: Cambridge University Press, 1974), 115.

portions, Shakespeare imagines unsettled, unstable, and uncertain relations among the parts of his created works.

Despite his apparent division of poetic labor, Puttenham does not leave the matter of "proportion" aside when he turns to *Ornament*, as he explains, "[F]igurative speaches [are] the instrument wherewith we burnish our language fashioning it to this or that measure or proportion" (143). They work by addition and subtraction, by degrees of "more" or "less": "[Figures are] alterations in shape, in sounde, and also in sence, sometime by way of surplusage, sometime by defect" (159). "Auricular" figures, those that affect the sound rather than the sense of language, proceed either by means of "amplification" or "abridgement": "[S]ometimes by *adding* sometimes by *rabbating* of a sillable or letter to or from a word" (161) and sometimes by adding or deleting whole words (for example, in *sillepsis* and *zeugma*, which translate in Puttenham's "Englished" version as "double supply" and "single supply," respectively). Many "sensable" figures too (those "altering and affecting the mynde by alteration of sence" [186]) are based on proportions, or deliberate disproportions, among words: in *sinathrismus* or the "heaping figure," for example, "we lay on such load and so go to it by heapes as if we would winne the game by multitude of words and speaches" (236). *Auxesis* is the figure of increase "because euery word that is spoken is one of more weight then another" (218), whereas *meiosis* works by a corresponding "diminution" (219).

Not all figures "amplify" or "abridge" the language, however. Others stipulate the quantitative relations among persons or ideas rather than among words. For example, *paragon* or *expeditio*, "the speedie dispatcher," works by proportioning degrees of human value:

> When a man wil seeme to make things appeare good or bad, or better or worse, or more or lesse excellent . . . then he sets the lesse by the greater, or the greater to the lesse, the equall to his equall, and by such confronting of them together, driues out the true ods that is betwixt them. (234)

Tropes that are based on identifications, such as metaphor, may thus be understood "mathematically" as well as visually: rather than "likenesses," they create "equalities." They are "sensable" analogues to auricular figures such as *parison* or "the figure of even," called so by Puttenham "because it goeth by clauses of egall length" (214). Proportionality in verse is a form of "analogy":

Now because this comelynesse resteth in the good conformitie of many things and their sundry circumstances, with respect one to another, so as there be found a iust correspondencie between them by this or that relation, the Greeks call it *Analogie* or a conuenient proportion. (262)

"Analogy" works by creating a "iust correspondencie" via "this or that relation": it's about the ways "things and their sundry circumstances" mean, or rather, are *made to mean*, with "respect one to another."

If these analogies do not hold, according to Puttenham, the poem will fail, as he observes: "It happens many times that to vrge and enforce the matter we speake of, we go still mounting by degrees and encreasing our speech with wordes or with sentences of more waight one then another." Among abuses of rhetoric, Puttenham includes *pleonasmus*, "too full speech" (257); *macrologia*, or "long language" ("when we vse large clauses or sentences more than is requisite to the matter" [257]); *tapinosis*, or "the Abbaser" ("It is no small fault in a maker to vse such wordes and termes as do diminish and abbase the matter he would seeme to set forth" [259]); and *bomphiologia* or "pompious speech" ("vsing such bombasted wordes as seeme altogether farced full of winde, being a great deale to high and loftie for the matter" [259–60]). Bad verse, in every case, is the consequence of the poet saying too much or too little in relation to "the matter." The ultimate standard of good verse is "decorum," or "the line and leuell for al good makers to do their busines by" (261); it is the highest principle, for Puttenham, of "keeping measure" (155). Above all, figuration is the art of "bewtifying [sentences] with a currant & pleasant numerositie" (160). It is no coincidence that Puttenham uses the same word to describe rhetorical effects as he earlier had metrical ones: "numerositie" is the primary cause, and the primary purpose, of poetry, the art of measuring words, in sound and in sense, as more, less, or equal to one another.

Shakespeare understands, with Puttenham, that poetic "numbers" may refer both to metrics and to "figures" of speech. In *Love's Labor's Lost*, Rosalind mocks the love poems sent her by her suitor Berowne:

Nay, I have verses too, I thank Berowne;
The numbers true, and, were the numbr'ng too,
I were the fairest goddess on the ground.
I am compar'd to twenty thousand fairs.
(5.2.34–37)

The phrase "numbers true" is a reference to Berowne's versification, but the second "numbr'ng" Rosalind cites is her "enumeration," that is, her appraisal in relation to "twenty thousand fairs." Such numbering is false, Rosalind implies, because the comparisons Berowne has made are overstated, disproportionate; for Shakespeare, comparisons fail not so much by kind as by degree. In sonnet 17, the poet expresses the wish that he could "in fresh numbers number all [the young man's] graces" (6). Here, again, Shakespeare conflates the counting of syllables ("fresh numbers") with the counting or enumeration of human value ("number all your graces").

To number love in numbers is the goal of Shakespeare's *Sonnets*. But it's not clear that the poet's desire is attainable: it depends on whether rhetorical numbers compose a fit rule for love.

"[His] worth's unknown, although his highth be taken" (sonnet 116.8): so Shakespeare's famous metaphor likens love to a star. Although its "highth" may be taken, its place in the sky observed and recorded, the value of the star remains unknown. Throughout his sonnet sequence, Shakespeare attempts to assess the "worth" of those he loves. For all the measuring he does, however, that worth remains uncertain, and Shakespeare comes to implicate poetic measures themselves in the making, rather than the meting, of human values.[4] If the worth of the young man cannot be known

---

[4] Joel Fineman's brilliant study of the *Sonnets, Shakespeare's Perjur'd Eye: The Invention of Poetic Subjectivity in the Sonnets* (Berkeley: University of California Press), elucidates the way Shakespeare reworked the terms of traditional epideictic poetry, or poetry of praise. For Fineman, Shakespearean praise is what happens, or what the poet believes is supposed to happen, when "mimesis and metaphor meet" (3). Fineman shows how the sonnets addressed to the "young man" are based on a "poetics of likeness," a metaphorics that works by means of "identificatory comparisons" (37); through "likeness," the poet assumes the power to reproduce the young man and to assimilate the young man to himself. In Fineman's deconstructive reading, the *Sonnets* reveal that reproductions are always marked by distance from an always already-deferred original, that representation can only be re-semblance. Mimesis breaks down as the poet's likenesses reveal difference, and the sequence ultimately defers to a full-scale "poetics of difference" in the poems addressed to the "dark lady" (188).

The terms of deconstructivist analysis, likeness and difference, however, may not be as apt for understanding Shakespeare's poetics as are the terms of "more" and "less"—where such degrees are not fixed but rather subject to continual reevaluation. Rather than the site where mimesis and metaphor meet, it may be more accurate to describe Shakespearean praise as the place where measure and metaphor meet: praise in the *Sonnets* is appraisal, the poet's efforts to assess worth by means of comparisons or rhetorical measures.

poetically, Shakespeare implies, it is because love is a dynamic and shifting *relation*, subject to the potential mismeasure of rhetorical figures.

That the poet is preoccupied with the *value* of his love(s), rather than simply their natures or kinds, is evident from the repetition of terms he uses to describe himself and those he desires. The poet speaks of their "worth," along with its variant forms (including "worthy," "worthiness," and "worthless") seventeen times in the *Sonnets*. He fears that in forty years' time the young man's beauty will be "of small *worth* held" (2.4); he contrasts the "inward *worth*" of the young man with his "outward fair" (16.11); he notes how other poets beside him "[s]peaking of *worth*, [failed to adequately mark] what *worth* in you did grow" (83.8); he tells the young man "your *worth*, wide as the ocean is" (80.5); he promises to continue "[p]raising [his] *worth*" despite time's cruel hand (60.14); he confesses that it was the dark lady's "*unworthiness* rais'd love in [him]" (150.13). Words in the same semantic field that occur frequently in the *Sonnets* include "merit" ("what *merit* lived in me" [72.2]; "place my *merit* in the eye of scorn" [88.2]; "[What] may express my love, or thy dear *merit*?" [108.4]; "What *merit* do I in myself respect?" [149.9]); "desert" or "deserving" ("the knowledge of mine own *desert*" [49.10]; "do more for me than mine own *desert*" [72.6]; "I should your great *deserts* repay" [117.2]; "your most high *deserts*" [17.2]; "how much more praise *deserv'd* thy beauty's use" [2.9]); and "esteem" and "estimate" ("'Tis better to be vile than vile esteemed" [121.1]).

Even when he appraises his worth at the highest rate, however, the poet wants "more" from the young man, as he makes clear from the opening sonnet of the sequence: "From fairest creatures we desire increase" (sonnet 1.1). The rhetorical terms of this desired "increase" are, often enough, mathematical. The young man must double himself (as opposed to staying "single" [3.14]); he must "live twice" (17.14); he must multiply himself by "ten times" (6.10); he must "sum [his] count" (2.11). Anticipating the gains the young man will accrue through sexual "use" or "usury," the poet hopes for a profit margin of "ten for one":

> That use is not forbidden usury,
> Which happies those that pay the willing loan;
> That's for thyself to breed another thee,
> Or ten times happier be it ten for one;
> Ten times thyself were happier than thou art,
> If ten of thine ten times refigur'd thee.
> (6.5–10)

The young man's children will "refigure" their father not just by resembling him but, according to the mathematical rhetoric of the poem, by "multiplying" him (Shakespeare's repetition of the word "ten" five times in five lines adds to the conceit of multiplication). The poet, in turn, produces "numbers" commensurate with the young man's:

> Be thou the tenth Muse, ten times more in worth
> Than those old nine which rhymers invoke,
> And he that calls on thee, let him bring forth
> Eternal numbers to outlive long date.
> (38.9–12)

The poet's "eternal numbers," ideally, will extend the young man's refiguration into infinity.

Yet the poet is often compelled to confess that his numbers either underestimate the young man or overestimate him. In sonnet 17, for example, the poet measures his verses against the young man's "deserts":

> Who will believe my verse in time to come
> If it were fill'd with your most high deserts?
> Though yet heaven knows it is but as a tomb
> Which hides your life, and shows not half your parts.
> (1–4)

Heaven is invoked as witness to the fact that his verses constitute less than "half" the parts of the young man. He fears that future readers may think his sonnets are written in "stretched metre" (12), that is, with syllables added to make up his lines. The poet urges the young man to have a child, that the "time to come" might believe his verses (17.1), since the young man would "live twice, in it [his child] and in my rhyme" (14). The problem is that the poet has already stated for the record (with heaven as witness) that his poem adds up to only "half" of the young man, so that even if his verses approximate his numbers, they can't (even with the addition of the child) add up to "twice" his value. Shakespeare's math here is deliberately inaccurate—that is, his numbers are "stretched"—in a vain effort to "count" living figures of the young man.

On the other hand, there may be "more" to the young man than any poet can express: "There lives *more* life in one of your fair eyes / Than both your poets can in praise devise" (83.13–14). Like "stretched meter," however, the "more" of praise is a rhetorical addition to the young man

that neither accounts for him accurately nor multiplies his deserts. As the
poet laments in sonnet 103:

> Alack, what poverty my Muse brings forth,
> That having such a scope to show her pride,
> The argument all bare is of more worth
> Than when it hath my added praise beside.
> (1–4)

In Puttenham's terms, the poet's appraisals represent an abridgment, or
an amplification, disproportionate to the young man. Indeed, the poet's
"added praise" may be the only "increase" he gains through his verses.

Shakespeare's poetic "enumerations" of the young man include his
comparisons, the relative measures that form the foundation of the art of
figuration. The most famous example in the sequence of a poem based
on a single, sustained figure of comparison, sonnet 18, compares the
young man favorably to a summer's day. The young man is declared at
once "more lovely" and "more temperate" (2): by comparison with a
summer's day, which "hath all too short a date" (4), the young man's
"summer" is an "eternal" one (9). But the young man's longevity owes
nothing to his native "temperament." Famously, it is not the young man's
intrinsic nature but the poet's own "rules" (his "eternal lines" [12], i.e.,
his verse) that make him "more" than a summer's day. Once again,
Shakespeare's poetic measures create, rather than find out, the desired
"addition" to the young man.

Above all, Shakespeare's rhetorical measures represent a will to proportion himself to his beloved, to correlate their desires. It is no wonder
that the word "part," as a term relating to proportionality, emerges as a
keyword of the sequence, occurring at least eighteen times. The poet
continually refers to the young man as a part of himself: "O, how thy
worth with manners may I sing, / When thou art all the better part of
me?" (39.1–2); "My spirit is thine, the better part of me" (74.8). He
speaks of "[t]hose parts of thee that the world's eye doth view" as against
the hidden parts that they "measure by thy deeds" (69.1, 10). As I have already mentioned, the poet fears that his verses show only half of the
"parts" of the young man, and he urges the young man to consider
the poet's parts in turn: "When thou reviewest this, thou dost review /
The very part was consecrate to thee" (74.5–6). The poet thrives on
whatever "parts" of the young man he may acquire:

> For whether beauty, birth, or wealth, or wit,
> Or any of these all, or all, or more,
> Intitled in [thy] parts do crowned sit,
> I make my love ingrafted to this store
>
> . . . .
>
> That I in thy abundance am suffic'd,
> And by a part of all thy glory live.
> (37.5–8, 11–12)

Despite the poet's claims of sufficiency in the young man's abundance, he ends, with deliberate ambiguity, as "a part of all"—sharing everything, perhaps, but more likely, accepting a far lesser "part" of him.

What makes it most difficult, perhaps even impossible, to gauge the relationship between the two men is the way that assessments of the poet's worth, and that of the young man, shift from poem to poem, from "measure" to "measure." In sonnet 87, the poet suggests that the young man's self-evaluation has changed over the course of their relationship:

> Farewell, thou art too dear for my possessing,
> And like enough thou know'st thy estimate;
>
> . . . .
>
> Thyself thou gav'st, thy own worth then not knowing.
> (1–2, 9)

The poet begins sonnet 62 with the highest "self-esteem," based on a sense of superior worth,

> Methinks no face so gracious as is mine,
> No shape so true, no truth of such account,
> And for myself mine own worth do define,
> As I all other in all worths surmount.
> (5–8)

Once again, however, the poet discovers he has miscalculated; he'd confused his own worth with that of the young man. Elsewhere he maintains that any account of his worth would be erroneous: "For you in me can nothing worthy prove; / Unless you would devise some virtuous lie, / To do more for me than mine own desert" (72.4–6)]. Still later, he defends his desert, accusing others of false appraisals and of projecting their own measures onto him:

> No, I am that I am, and they that level
> At my abuses reckon up their own;
> I may be straight though they themselves be bevel.
> (121.9–11).

No wonder "'[t]is better to be vile than vile *esteemed*" (121.1) in the *Sonnets*, when varied and vacillating measures replace the question of what one is for how one, in any given moment, may be valued.

For the poet to achieve a perfect one-to-one ratio with his beloved, the poet's "numbers" ought to "equal" the young man's. In sonnet 84, however, the young man is declared a nonpareil; no poetic comparison is enough to make up his "equal":

> Who is it that says most, which can say more
> Than this rich praise, that you alone are you,
> In whose confine immured is the store
> Which should example where your equal grew?
> (1–4)

The poet who says, merely, "You are you" is the one who best "dignifies" (line 8; literally, makes worthy) his verse by creating a perfect paragon or "counterpart" to the young man. But if the young man (in this particular assessment) has no equal, if he can't be compared to anyone, or anything, he can't, by Shakespeare's numbers, be counted at all. As "one" ("you alone are you"), the young man is "nothing," for according to a Renaissance proverb, "One is no number." The poet makes explicit reference to this notion in sonnet 136:

> In things of great receipt with ease we prove
> *Among a number one is reckon'd none:*
> Then in that number let me pass untold,
> Though in thy store's account I one must be,
> For nothing hold me, so it please thee hold
> That nothing me, a something sweet to thee.
> (7–12, emphasis added)

The Renaissance notion that one equals zero is a mathematical one. What we call the number one is a "unit" in early modern books on arithmetic: "Unitie, unit, or one is, by which euerie thing that is, is sayd one," while "Number is, a multitude or a many of units."[5] The proverb "One is no number" derived from the Pythagorean idea that One, as the antithesis of Many, cannot itself be subdivided; it cannot be a number because it contains no plurality.[6] Thus in sonnet 136, the poet "proves," through mathematical reasoning, that as "one" (that is, as "nothing"), he need not be "accounted." It is the same mathematical concept that underlies the final warning of sonnet 8, "Thou single wilt prove none" (14),[7] and the joke in which Nature, in sonnet 20, turns his lover from a woman to man, adding "one thing to [the poet's] purpose nothing" (12). If the young man is no number, he cannot be proportionately "numbered" by the poet's verses—a crisis for love and for art. Indeed, in reminding us that "One is no number," the poet goes further, suggesting that mathematics is not an appropriate measure for understanding the "one" he loves. The problem is that without "enumeration," there is no way for the poet to find out the relation of love, to determine the intimacy or the distance between them and say, with any accuracy, "Thus far the miles are measur'd from [my] friend" (sonnet 50.4).

Although the poet claims to "need [no] tallies [the young man's] dear love to score" (sonnet 122.10), his rhetorical numbers attempt to be just that: a method for counting or "tallying" his love. He suggests that absence from his beloved has taught him "how to make one twain" (39.13), but that's one of many lessons in arithmetic he never really masters. Despite the poet's efforts, he cannot make "one" into "two," neither by multiplying the young man nor by dividing him; nor can he make "two" into "one." In the mathematical terms he invokes throughout the sequence, the young man remains "nothing" to him. If he seems to arrive, at times, at answers to the question of their relation, the answers are only those that his numbers themselves prefigure. The poet himself is the (mis)measure of his beloved; he is the source of "false esteem" (sonnet 127.12) in the *Sonnets*. Although his "highth" is taken throughout the

---

[5] Thomas Masterson, *First Booke of Arithmeticke* (London, 1592), 1.

[6] Karl Menninger, *Number Words and Number Symbols: A Cultural History of Numbers*, trans. Paul Broneer (Cambridge, MA: MIT Press, 1969), 19.

[7] Stephen Booth notes some of Shakespeare's allusions to this proverb in his edition of Shakespeare's *Sonnets* (New Haven: Yale University Press, 1977).

sequence, the worth of the young man remains incalculable. By the end
of his "numbered" sequence of sonnets, it becomes tragically clear that
only Shakespeare's own figures, his own enumerations, can be counted,
and recounted, again and again. The measures, themselves, are all.

I have already argued that Puttenham's book on *Ornament* is also, like his
book on meter and stanza form, centered on questions of proportionality, the proper relationships among the parts of a verse. The art of poetry
is, all told, an art of keeping measure. Puttenham's book *Of Proportion*,
however, makes special claim to problems of literary measurement; it
deals explicitly with the "numbers" of verse, with what Hamlet refers to
when he speaks of his efforts to "reckon" his groans. In his chapter "Of
Proportion in Measure," Puttenham explains that "[m]eter and measure
is all one, for what the Greekes called *metron*, the Latines call *Mensura*,
and is but the quantitie of a verse, either long or short" (67). In the remainder of this chapter, I will reconsider the long-standing critical problem of measuring Shakespeare's verses, focusing especially on how
Shakespeare himself might have scanned his own lines. Evaluating
Shakespeare's metrics within its historical context, I will show how taking Shakespeare's meters on his own terms may alter our perception of
what we think we hear in his works. Above all, I will revisit the commonplace that Shakespeare developed what we might call a "human metrics," a "natural" measure of human character by the rhythms of verse.
    As Puttenham and his contemporaries would have readily conceded,
the determination of the "quantitie" of a verse was not always self-
evident. It depended, first of all, on what poets were choosing to count in
the constitution of their lines. Whereas ancient Greek and Latin poets
measured the "length" of syllables, that is, their quantity as duration in
time, vernacular poets counted the number or "quantity" of syllables per
line. Philip Sidney summarized the two metrical systems, "the one Auncient, the other Moderne: the Auncient marked the quantitie of each sil-
able, and according to that framed his verse; the Moderne observing
onely number. . . . Those rimes we commonly use observing nothing but
the number of sillabes."[8] Although Sidney (like Puttenham) finds much
to admire in native verses, his language here betrays a dominant bias of
the period toward classical metrics: "we" observe "*onely* number," "*noth-*

[8] Sir Phillip Sidney, *An Apologie for Poetrie*, in *Elizabethan Critical Essays*, 2 vols., ed. G.
Gregory Smith (Oxford: Oxford University Press, 1904), 1:148–207, quote at 204.

ing but the number of sillabes." As is well known, even Renaissance apologists for English poetry feared that the "numbers" of vernacular verses were somehow inadequate, inferior to those of Greek and Latin.

Puttenham's *Of Proportion* is in fact a response to a belief, common enough in the sixteenth century, that English poetry had no meter, no "true" numbers at all, and moreover that the English language itself was intrinsically unfit for true measure. From William Lily's official school grammar, followed by years of exercises in Latin prosody, Puttenham's generation learned that versification was the art of arranging syllables, defined according to "quantity" or duration, in patterns of "longs" and "shorts." Syllable length was understood as a matter of time—how long it actually took to pronounce each syllable—with a long syllable theoretically taking twice the time of a short syllable. To create classical verses, poets arranged sequences of syllables to create lines that were perceived as quantitatively equivalent. The lines might be constituted of identical sequences of syllables (i.e., sequences with the same number of longs and shorts arranged in the same way). Alternatively, sequences of syllables might be deemed equivalent if they were of equal duration, for example, in the case of the relationship between two short syllables with one long. In classical prosody, sequences containing the same number of syllables might also be deemed equal (Renaissance theorists, who think of syllable count as the concern of vernacular poets only, do not tend to acknowledge this). As one scholar of classical metrics sums up the possibilities: "These equivalences involve the working of three different rhythmic principles which can be called, respectively, the principle of identity, the principle of quantitative equality, and the principle of syllabic equality."[9] For Renaissance English students and practitioners of versification, however, the two former principles were paramount, and both marked time or duration as the salient feature of "quantified" verse.

Yet as Derek Attridge has discussed in his definitive study of sixteenth-century attempts to introduce classical meters into English, the rules for "duration" were applied inconsistently, even arbitrarily at times, in Latin, and later in English verse as well. For example, although Latin consonants were held to add to syllable length, those positioned at the onset of a syllable were not counted, while long vowels or diphthongs followed by two or more consonants were not judged to be any "longer"

[9] Thomas A. Cole, "Classical Greek and Latin," in *Versification: Major Language Types*, ed. W. K. Wimsatt (New York: New York University Press, 1972), 66.

than a long vowel or diphthong alone (for example, in *mens* versus *mē*).[10]
Such disparities, Attridge explains, came as the result of a system of rules
surviving dramatic changes in the language: they may have once per-
tained to Greek and perhaps to early Latin, but "the quantities on which
Latin verse was based ceased [by the fifth century] to be a property of
the spoken language and had to be learned for the purpose of scanning
and writing poetry in classical metres." Long before Renaissance poets
and poetic theorists had inherited it, Latin prosody was an abstract system, complete, fixed, and incommensurable with "true" properties of
the language.

"Quantity" was not only understood literally to be present in Latin syllables, however; it was generally assumed to be a universal feature of syllables across languages.[11] Sixteenth-century poets and poetic theorists
thought they actually "heard" it not only in Latin but in English as well.
George Gascoigne claimed to hear quantitative meters in Chaucer:

> [O]ur father *Chaucer* hath used the same libertie in feete and
> measures that the Latinists do use: and who so euer do peruse and
> well consider his workes, he shall finde that although his lines are
> not alwayes of one selfe same number of Syllables, yet, beyng
> redde by one that hath vnderstanding, the longest verse, and that
> which hath the most Syllables in it, will fall (to the eare) corre-
> spondent vnto that whiche hath the fewest sillables in it.[12]

In other words, Gascoigne believed that Chaucer was attempting to proportion his lines by creating lines of equivalent duration, despite uneven
syllable counts, via combinations of long and short feet. For Gascoigne,
however, as for so many others, as Attridge makes clear, "the ear allowed
itself to be governed by the mind."[13]

Most agreed, and still agree, that Chaucer was in fact attempting to
write lines of "one selfe same number of Syllables," that is, to use syllabic
prosody in the manner of medieval continental poetry. French decasyllabic verse, originating in the eleventh century, is generally held to be

[10] Attridge, *Well-Weigh'd Syllables*, 10.
[11] Ibid., 21, 119.
[12] George Gascoigne, *Certayne Notes of Instruction*, in Smith, *Elizabethan Critical Essays*,
1:46–57, quote at 50.
[13] Derek Attridge, *The Rhythms of English Poetry* (New York: Longman, 1982), 16.

Chaucer's primary model for his own metrics, based on lines of ten syllables. Along with Italian syllabic prosody, especially the hendecasyllabic
verse of Petrarch, Chaucer in turn provided Renaissance English writers
with a primary model for vernacular verse. It should be noted that alongside its "numbers" (i.e., the counting of syllables), rhyme was, for Renaissance writers, specific to romance verse. Indeed, rhyme was so conspicu-
ous a feature of vernacular poetry that the word "rhyme" or "rime" was
often used as shorthand for the whole prosodic system.

Given the prestige of Latin and Latin versification, it is not surprising
that so many Renaissance English authors hoped to find a way to "quantify" English metrics. Roger Ascham called for English poets to find "iust
measure in euerie meter, as euery ignorant person may easely do, but
also trew quantitie in every foote and sillable, as onelie the learned
shalbe able to do," mocking "rhymers" or "rash ignorant heads, which no
can easely recken up fourten sillabes."[14] Thomas Campion, chief among
advocates of introducing quantitative measures into English, argued
that rhyme "forestall[s] the right of true numbers." For Campion as for
Ascham, "truth" in number depended on proportionality:

> Number is *discreta quantitas*: so that when we speake simply of
> number, we intend only the disseuer'd quantity; but when we
> speake of a Poeme written in number, we consider not only the
> distinct number of the sillables, but also their value, which is con-
> tained in the length or shortnes of their sound . . . so in a verse the
> numeration of the sillables is not so much to be obserued as their
> waite and due proportion.[15]

"Proportion" depends not on the arithmetic "counting" of syllables but
on a determination of "values," on proportional "equations" made
among words; as Campion complains, "in our kind of riming what proportion is there kept where there remaines such a confused inequalities
of sillables?"[16] Even Sidney, who was far less hostile to "riming," cannot
help but admire those classical poets who, "peyzing each sillable of each
word," discovered "iust proportion" in verse.[17] Those who imagined that

14 Roger Ascham, *Of Imitation*, in Smith, *Elizabethan Critical Essays*, 1:1–45, quote at 31.
15 Thomas Campion, *Observations in the Art of English Poesie*, in Smith, *Elizabethan Critical Essays*, 2:327–55, quote at 328.
16 Ibid., 330
17 Sidney, *Apologie for Poetrie*, 160.

verse form might typify the divine plan seem to have been thinking specifically of quantitative verses.[18] For many Renaissance English writers, "true" number remained an instrument of *relation*, of distributed values rather than fixed sums; simply adding up the syllables of a line didn't count.

Samuel Daniel is often credited with turning the tide of vernacular metrical theory in favor of native systems of measure. His *Defence of Ryme* (1603) counters the claim that syllabic verse is "grosse, vulgare, barbarous" by appealing to its "fitness" for the vernaculars:

> Euery language hath her proper number or measure fitted to vse and delight, which Custome, intertaininge by the allowance of the Eare, doth indenize and make naturall. All verse is but a frame of wordes confined within certaine measure, differing from the ordinarie speach, and introduced, the better to expresse mens conceipts, both for delight and memorie. Which frame of words consisting of *Rithmus* or *Metrum*, Number or measure, are disposed into diuers fashions.

Daniel concedes that all meters vary from "ordinarie speech" by means of a "certaine measure" that confines it. Yet every language has a "proper number or measure" that best suits it; this measure may not be intrinsically natural but comes to seem so by customary use. Rhyme itself, for Daniel, is a "[h]armonie farre happier than any proportion Antiquitiie could euer shew vs," creating a "cloze" or formal limit to the unbounded feelings that motivate the poem:

> For the body of our imagination being as an vnformed *Chaos* without fashion, without day, if by the diuine power of the spirit it be wrought into an Orbe of order and forme, is it not more pleasing to Nature, that desires a certaintie and comports not with that which is infinite, to haue these clozes, rather than not to know where to end, or how farre to goe, especially seeing our passions are often without measure?[19]

---

[18] Attridge, *Well-Weigh'd Syllables*, 115.
[19] Samuel Daniel, *A Defence of Ryme*, in Smith, *Elizabethan Critical Essays*, 2:356–84, quotes at 357, 359, 366.

Daniel admits that our passions themselves are beyond measure and that our ideas represent an "vnformed Chaos without fashion," "that which is infinite." Poets fashion measures that depart from Nature and yet provide a form necessary for its expression.

Daniel, in defending "rhyme" or native versification, offers an intervention in an ancient debate concerning whether "meters" are natural to poetry or somehow extrinsic to it. Aristotle in his *Poetics* had said that the poet should be a maker of plots [mythoi] rather than verses [metron], a view echoed by Sidney, who wrote that verse was "but an ornament but no cause to Poetry."[20] Julius Caesar Scaliger was alone among major Renaissance theorists in explicitly countering this idea; he pronounced that the poet was above all a maker of *verses*.[21] Daniel thus offers a compromise in the debate over the "nature" of meters. All measures of all kinds, Daniel suggests, should ideally correspond to the objects that they measure. He is not alone among his contemporaries in his belief that the human body provides the best example of a "natural" measure: "[T]he best measure of man is to be taken by his owne foote bearing euer the neerest proportion to himselfe, and is neuer so farre different and vnequall in his powers, that he hath all in perfection at one time." For Daniel, "ryme" and syllabic prosody are for English the measure that conforms most closely to the nature of the language. He tries to put the ghost of classical quantification to rest: "And as for those imagined quantities of sillables, which haue bin euer held free and indifferent in our language, who can inforce vs to take knowledge of them?" If English poets pledge allegiance to native "rule," they must stop measuring their verses "by the square of Greece and Italie."[22]

It is generally believed that Shakespeare, in his own metrics, anticipates in practice what Daniel elucidates in theory: Shakespeare reveals the "natural" character of English syllabic prosody and its commensuration with human nature. I suggest, however, that for Shakespeare, as for so many of his contemporaries, the prestige of classical metrics (especially, its emphasis on the "value" rather than just the "number" of sounds) still held sway. Contrary to the ways his verses are generally mea-

---

[20] Sidney, *Apology for Poetrie*, 159–60.
[21] O. B. Hardison, *Prosody and Purpose in the English Renaissance* (Baltimore: Johns Hopkins University Press, 1989), 68.
[22] Daniel, *Defense of Ryme*, 370–71, 378, 364.

sured, I will show how Shakespeare attempted, in a limited but significant way, to re-create vernacular numbers as a proportional art, one that
worked by contrast and comparison, by relative "weights" rather than
merely by counting. He was certainly not "peyzing each syllable" as long
or short, but his mature verse achieves effects that can only be understood, in the context of Renaissance metrical theory, as quantitative. We
may have mismeasured the aims and effects of Shakespeare's poetic
"reckonings."

Ever since Ben Jonson compared Shakespeare's verses favorably to "Marlowe's mighty line," pronouncing that Shakespeare's "*living* line[s]" have
the approval of Nature herself ("Nature herself was proud of his designs
/ And joyed to wear the dressing of his lines"),[23] Shakespeare's meters
have been heard as "natural" ones. This is especially true of his dramatic
meters and even more of the meters of his later plays. O. B. Hardison, for
example, finds Shakespeare partial to "showy" or otherwise formal verse
forms in his early plays, but "in the mature plays the norm is closer to natural speech."[24] For Susanne Woods, Shakespeare's dramatic meters become so natural as to be "mimetic of character, and of changes in character."[25] George Wright does not agree that Shakespeare's characters are
distinguished by their meters or that particular characters have distinctive "metrical destinies" but agrees with Woods that Shakespeare's meters are "mimetic" of psychological and emotional states of mind.[26] All
concur that Shakespeare's meters come to echo human personality and
human thought, that they become "lines of life" (sonnet 16.9), "natural"
measures of man.

---

[23] Ben Jonson, "To the memory of my beloved, The Avthor, Mr. William Shakespeare," dedicatory poem to the First Folio of 1623, printed in William Shakespeare,
The Riverside Shakespeare, 2nd ed., ed. Blakemore Evans (Boston: Houghton Mifflin,
1997).
[24] Hardison, *Prosody and Purpose*, 254.
[25] Suzanne Woods, *Natural Emphasis: English Versification from Chaucer to Dryden* (San
Marino: Huntington Library, 1984), 246.
[26] George T. Wright, *Shakespeare's Metrical Art* (Berkeley: University of California
Press, 1988), 254. I have quoted Wright extensively throughout this chapter because I
hold his book to be the finest on the subject of Shakespeare's meters; his chapters are
filled with sensitive and persuasive readings. I am not convinced, however, by his emphasis on "feet" as the best way (or the only way) to explain the metrical effects of
Shakespeare's lines.

We must now reconsider some of the assumptions behind the long-standing idea that Shakespeare discovered, or invented, a "natural" English metrics. In the late plays, for example, what sometimes passes as *natural meter* are verses that are, more precisely, *less metered* (there is an apparently slight but in fact essential distinction here). Moreover, the suggestion that Shakespeare heard certain meters as more natural than others, in *kind*, is a misleading one. The way Shakespeare's meters work—the aesthetic and thematic effects they produce—has nothing to do with the inherent qualities of particular measures but rather with the way those measures are set against one another, the way they "relate" in particular dramatic contexts. If Shakespeare's meters seem to "mean" something, it is in large part because he proportions his words in contingent relations to one another, "weighs" them according to changing criteria of poetic value. And in doing so, crucially, Shakespeare most likely saw the metrics of his late plays as far more "artful" than natural.

When metricists today listen to Shakespeare's numbers, most agree that they hear iambic pentameter, rhymed in the poems and, more often than not, unrhymed in the plays. Although the rise of unrhymed iambic pentameter, or blank verse, is generally credited to Christopher Marlowe (the phrase "blank verse" was coined by Thomas Nashe in 1589), Shakespeare is hailed as the poet who discovered, and then exploited, its truly "English" nature. Modern metricists often cite the "naturalness" of iambic pentameter itself, at least in English: "Iambic pentameter has often been called the most speechlike of English meters, and this is undoubtedly true, especially of its blank verse form." But even critics who hear blank verse as the most "natural" of poetic meters sometimes hedge on the question of whether iambic pentameter is natural to the language itself: "Whether it is true because English is a naturally iambic language is a more questionable claim."[27]

It is essential to remember that the assessment of iambic pentameter as the most "natural" of meters—with Shakespeare's blank verse as chief exemplar—is a modern one. Renaissance poets were not so sure. Non-

---

[27] Ibid., 1. Unwilling to commit to the idea that "iambic pentameter" is "really" natural, Wright concludes that "iambic pentameter [in Shakespeare's plays] should be taken as a kind of figure for natural speech." He suggests that English "speech lends itself to the illusions that it is iambic, that life might actually be lived in lines" (189).

dramatic blank verse was introduced to England by Henry Howard, Earl of Surrey, in his translation of books 2 and 4 of the *Aeneid*, as an alternative to English "rhymes" and an analogue to classical heroic meters (which were also unrhymed). As late as the following century, Milton's choice of blank verse as his epic meter was a deliberate effort to recall classical versification; he called blank verse an "ancient liberty recover'd . . . from the troublesome and modern bondage of Riming."[28] In nondramatic writing of the period, blank verse was not intended to sound "natural" or speechlike at all; it was heard as something more distant, more dignified, an echo of the classical high style. Indeed, blank verse signaled precisely a departure from native forms.

As a vehicle for drama, blank verse was introduced by Thomas Sackville and Thomas Norton's *Gorboduc* (1561). Their usage was, like Surrey's, an effort to give the play an elevated, foreign character. They were seeking a sound appropriate to Senecan tragedy (the contemporary Italian tragedians they were imitating wrote Senecan plays in Italian blank verse). The title page of the second edition of Surrey's translation makes explicit reference, in fact, to its "straunge meter."[29] Tucker Brooke sums up the early uses of blank verse on the Renaissance stage: "It was consciously as a strange or foreign metre that blank verse was employed in England throughout the period before Marlowe. It seems to have been valued chiefly as a proper means of translating or simulating the exotic grace of Latin quantitative verse."[30]

It is not surprising, then, that so many contemporary references to dramatic blank verse tended to emphasize not its "natural" or "speechlike" qualities but its "inflated" sound. Thomas Nashe, in his preface to Robert Greene's *Menaphon*, berates those dramatists who

> intrude themselues to our eares as the alcumists of eloquence, who (mounted on the stage of arrogance) think to outbraue better pens with the swelling bumbast of a bragging blanke verse. Indeed, it may be the ingrafted ouerflow of some kilcow conceipt,

---

[28] John Milton, "The Verse" in *Paradise Lost*, in *John Milton: Complete Poems and Major Prose*, ed. Merritt Y. Hughes (New York: Macmillan, 1957).
[29] See Hardison's discussion of what might have been considered "strange" about Surrey's meters (*Prosody and Purpose*, 130–32).
[30] Tucker Brooke, "Marlowe's Versification and Style," *Studies in Philology* 19 (1922): 186–205, quote at 187–88.

that ouercloieth their imagination with a more than drunken resolution, beeing not extemporall in the inuention of anie other
meanes to vent their manhood, commits the digestion of their
cholerick incumbrances to the spacious volubilitie of a drumming
decasillabon.[31]

Blank verse, according to Nashe, "intrudes" on English ears; it is not
"natural" but "swelling," excessive in its "volubilitie." Nashe was by no
means the only contemporary writer to identify blank verse with "bombast" (literally, cotton stuffing). The word was frequently used in reference to foreign or showy language, as these contemporary citations from
the *OED* attest: "Then strives he to bumbast his feeble lines / With farre-
fetcht phrase" (1599); "That doth . . . bumbast his labours with high
swelling and heaven-disimbowelling words" (1603); "A bumbast circumstance / Horribly stufft with Epithites of warre" (1604).[32] Joseph Hall
hears Marlowe's dramatic language as "pure Iambick verse," but he
marks that verse as distinctively foreign:

> [If the author] can with termes Italianate,
> Big-sounding sentences, and words of state,
> Faire patch me up his pure Iambick verse,
> He ravished the gazing Scaffolders.

Although Ben Jonson later celebrated Marlowe's line as "mighty," he also
deemed it strained and unnatural:

> The true Artificer will not run away from nature, as hee were
> afraid of her, or depart from life and the likenesse of Truth, but
> speake to the capacity of his hearers. And though his language dif-
> fer from the vulgar somewhat, it shall not fly from all humanity,
> with the *Tamerlanes* and *Tamer-Chams* of the late Age, which had
> nothing in them but the *scenicall* strutting and furious vociferra-
> tion to warrant them to the ignorant gapers.

Marlowe, however, was not alone in receiving this censure. At least one
of his contemporaries thought Shakespeare just as "stuffed": Robert

---

[31] Thomas Nashe, preface to Greene's *Menaphon*, in Smith, *Elizabethan Critical Essays*,
1:307–20, quote at 308.
[32] *Oxford English Dictionary*, s.v. "bombast."

Greene charges Shakespeare with "suppos[ing] he is as well able to bombast out a blank verse as the best of you."[33]

In fact, the Renaissance critique of blank verse as "bombast" cut two ways. It was, on the one hand, a way of ridiculing lines that seemed artificially inflated in (sometimes bad) imitation of classical heroic poetry. On the other, however, it was a way of jeering all over again at native "numbers," not at "iambic" verses per se but rather at the practice of counting out syllables, of artlessly adding (or deleting) sounds simply to make up the line. Nashe elsewhere mocked those who "bodge vp a blanke verse with ifs and ands," poetasters who forego "true" verse in favor of an artificially "bombasted" one.[34] When Shakespeare's dramatic characters make reference to "blank verse," they often do so derisively. Orlando in *As You Like It* calls out to Rosalind: "Good day and happiness, dear Rosalind!" and Jacques snidely replies, "Nay then God buy you, and [i.e., if] you talk in blank verse" (4.1.30–32). Hamlet laughs about an imagined performance by a player, "[T]he lady shall say her mind freely, or the blank verse shall halt for 't" (*Hamlet*, 2.2.324–25); he seems to mean that the blank verse will "stumble" if the actor omits any (indecent) words. It was as a synonym for "numbers" or native, syllabic prosody that blank verse came to be heard as a vernacular verse form rather than as an analogue to a foreign one.

To the extent that blank verse was considered synonymous with "iambic" poetry, Renaissance writers had another reason to identify it with "common" language. Although Horace claimed that iambics were suited to the representation of action, Aristotle had likened them to everyday speech.[35] Campion, in an effort to reconcile these positions, made a distinction between "pure Iambick," which he considered appropriate to "the Tragick and Heroick Poeme" (i.e., as an analogue to Latin dactylic hexameters), and what he termed the "licentiate iambic": "The Iambick verse in like manner being yet made a little more licentiate, that it may therby the neerer imitate our common talke, will excellently serve for Comedies."[36] What made "licentiate" iambics sound more like "common talke," for Campion, was, however, its flexibility, its tendency to break metrical rules. In other words, licentiate iambics tended toward

---

[33] Hall, Jonson, and Greene as quoted in Hardison, *Prosody and Purpose*, 236–37.
[34] Nashe, preface to Greene's *Menaphon*, 312.
[35] Hardison, "Crosscurrents," 123.
[36] Campion, *Observations*, 338.

prose. Sixteenth-century Italians began to experiment with prose as a vehicle for drama, and in 1566 Gascoigne produced an English prose comedy, *Supposes*, a translation of Ariosto's *I suppositi*. By the late sixteenth century, prose had become a conventional vehicle for the speech of comic and lower-class characters in English drama as well.[37] Renaissance writers who confused iambic meter with prose may have furthered its gradual identification with "natural" speech.

One thing, so far, should be clear: standards of "naturalness" shifted dramatically over the course of the sixteenth century and varied according to the tastes of individual poets and poetic theorists. After all, in the 1560s fourteeners were accepted as a way of representing "real" speech.[38] Even Daniel, as prudent and practical as any metricist of the period, had his idiosyncrasies: while he argued the natural qualities of rhyme, he excluded couplet rhymes, calling for their banishment from English poetry.[39] Where meter was concerned, the ear was continually assimilating itself to a changing mind.

Today, "blank verse" is understood to mean not only unrhymed, syllabic verse but, more specifically, unrhymed iambic pentameter. It is generally agreed that Renaissance writers understood it that way too. If they did, it means they believed that lines of English verse were or could be measured in "feet," units of two or three syllables, whether iambic, trochaic, dactylic, or any of the others prescribed by the ancients and arranged in patterns. Renaissance writers inherited the notion of the foot as the basic unit of verse measurement, along with the system for metering lines by combining feet of varying lengths, from classical metrics; once again, the basic text for the transmission of ideas about the nature and varieties of poetic feet was Lily's grammar. Gascoigne is generally credited with establishing the foot as the basis for scanning English lines, although its aptness for vernacular poetics remained controversial throughout the period. Gascoigne believed that native poetry only made use of *iambic* feet: "[We] vse none other order but a foote of two sillables, whereof the first is depressed or made short, and the second is eleuate and made long." For Gascoigne, however, one type of foot did not a verse make: "And sherely I can lament that wee are fallen into suche a playne and

---

[37] Hardison, "Crosscurrents," 125.
[38] Hardison, *Prosody and Purpose*, 253.
[39] Daniel, *Defense of Ryme*, 382.

simple manner of wryting, that there is none other foote vsed but one;
wherby our Poemes may iustly be called Rithmes, and cannot by any
right challenge the name of a Verse." There is already confusion here, of
course: Gascoigne is imagining he hears syllable length—long versus
short syllables—when what he's talking about may well be stressed and
unstressed syllables. (For all his emphasis on iambs, however, he defines
the sonnet as consisting of verses "of fourteen lynes, euery line conteyn-
ing tenne syllables," and never mentions feet at all.)[40] Thomas Campion
concurred with Gascoigne's sense that English lends itself to iambics; in
fact Campion finds that iambic feet "fall out so naturally in our toong,
that, if we examine our owne writers, we shall find they vnawares hit of-
tentimes vpon the true Iambick numbers."[41] Campion also insists, however, that "iambic," even in English, refers to a pattern of short and long
syllables, defined by quantity or duration in time.[42]

Puttenham, on the other hand, does not allow for the existence of feet
in English verse. Because English syllables, in his view, do not observe duration, they cannot be measured by the terms of classical quantification.
Distinguishing classical verse from English rhymes, he notes that "[t]his
quantitie with them [Greek and Latin poets] consisteth in the number of
their feete: and with vs in the number of sillables" (67). Poetic feet, he reminds us, depend on time, the "pace" at which verses run:

> [Y]e may say, we haue feete in our vulgare rymes, but that is im-
> properly: for a foote by his sence naturall is a member of office
> and function, and serueth to three purposes, that is to say, to go,
> to runne, and to stand still: so as he must be sometimes swift,
> sometimes slow, sometime vnegally marching or peraduenture

---

[40] Gascoigne, *Certayn Notes*, 50, 55.

[41] Campion, *Observations*, 333.

[42] It is possible, as Hardison notes, that Renaissance poets and poetic theorists such
as Campion were really hearing stress when they claimed to be hearing "number"
(*Prosody and Purpose*, 18). Indeed, the theory that Shakespeare was writing "iambic
pentameter" is based on this premise. I have not addressed here either the history or
the practice of measuring accent or stress in English verse—for example, the merits
of describing Shakespeare's line as "accentual-syllabic." See Woods for an excellent
discussion of the terms *accent*, *stress*, and *ictus* in metrical theory (*Natural Emphasis*, 5).
I have generally avoided modern metrical vocabulary here in my effort to explore
Shakespeare's own terms of measure; i.e., since Shakespeare didn't speak of "stress," I
have chosen not to, either.

steddy. And if your feete Poetically want these qualities it can not be sayd a foote. (67)

Daniel agrees with Puttenham and goes further: in contrast to Campion especially, Daniel deems "foot" meter, based on quantity, as utterly foreign to the nature of the vernacular. What some call English "iambics," Daniel insists, are really decasyllabic verses under a new (and inappropriate) name:

> [W]hat strange precepts of Arte about the framing of an Iambique verse in our language? which, when all is done, reaches not by a foote, but falleth our to be the plaine ancient verse, consisting of ten sillables . . . which hath euer beene vsed amongest vs time out of minde, and, for all this cunning and counterfeit name, can or will [not] be any other in nature then it hath beene euer heretofore.[43]

Wright concludes sensibly that even if some theorists, particularly Gascoigne, were aiming toward the modern idea of English iambics—a line consisting of pairs of one unstressed and one stressed syllable—the classical identification of poetic feet with syllable "length" utterly confused the terminology of vernacular metrics, and with it, the measurement of English poetry. All that Renaissance writers were certain of was that vernacular lines were decasyllabic; in fact it is possible that "for Elizabethan readers . . . iambic [meter] was perceived essentially as a line whose pattern was entirely defined when you stated that it had ten syllables." Out of respect for Renaissance terminology, some recent metricists, in fact, have followed suit in preferring to call the dominant line of Renaissance verse "decasyllabic" rather than "iambic."[44]

In the history of English metrics since the Renaissance, as it happens, no concept has proven more enduring, or more controversial, than the notion of the foot. Some scholars hold "the emergence of foot meter as central to the development of English verse in the Renaissance."[45] Others, however, argue that the foot was and remains a convention, an "in-

---

43 Daniel, *Defense of Ryme*, 376–77.
44 Wright, *Shakespeare's Metrical Art*, 39, 19.
45 Woods, *Natural Emphasis*, 5.

denized" custom that bears no intrinsic relation to the nature of the language: "One could choose almost any foot for the basic metre and explain the actual pattern of stresses in terms of substitutions; indeed, one of the weaknesses of the classical approach is that any succession of syllables can be divided into recognized feet. But the choice of a basic foot here would be an arbitrary one, not reflecting anything in the reader's experience."[46] Even those who "hear" feet in English are often obliged to admit that any "two experienced scholars of versification" may disagree about the scansion of Renaissance English lines.[47] It is the measurement system itself, according to Attridge, not the English language, that "invite[s] some audible manifestation of the ghostly divisions on which it is based, and [makes] phonetic equivalences which are no more than theoretical." Linguists who work on Renaissance meters have analogous disagreements about the reality of poetic feet and their commensurability with the nature of English.[48] Although the classical, quantified foot has generally been rejected as a suitable measure for English, it is telling that we still use the word "foot" nonetheless: our terminology has reified as "traditional" these early irregularities of metrical analysis. Perhaps the anthropometric term "foot," which naturalizes the measure rhetorically, has something to do with its hold on our poetic imaginations.

If the testimony of sixteenth-century writers suggests divergent ways of assessing blank verse in the Renaissance, how did Shakespeare himself conceive them? Did he consciously write in "iambic pentameter"? Did he intend to create a "natural" meter in his poems and plays? Given the confusion and lack of consistency in the terminology and theory of English metrics in the period, it is risky to assume that Shakespeare heard what we think we hear in his verses.

[46] Attridge, *Rhythms of English Poetry*, 12.

[47] Woods, *Natural Emphasis*, 10–11.

[48] For the best discussion of English metrics from a linguistic perspective, see Paul Kiparsky, "The Rhythms of English Verse," *Linguistic Inquiry* 8 (1977): 189–247. Elaborating and refining Kiparsky's system for the scansion of English poetry, Kristin Hanson has proposed a system for understanding iambic pentameter that may well put the idea of poetic "feet" to rest once and for all. As an analogue to parameter-based theories of syntax, she theorizes a set of parameters that condition and limit the metricality of lines within a basic template of alternating weak and strong positions in a line. See Kristin Hanson, "From Dante to Pinsky: A Theoretical Perspective on the History of the Modern English Iambic Pentameter," *Rivista di Linguistica* 9.1 (1996): 53–97; and her forthcoming book on English metrics.

There *is* clear, incontrovertible evidence that Shakespeare deliberately and carefully counted syllables in his *Sonnets*. Like Holofernes in *Love's Labor's Lost*, who complains that Nathaniel "find[s] not the apostraphas" in his verse (4.2.119), Shakespeare seems to have exploited all available resources in early modern English for contracting and expanding the syllables of words to suit his "numbers." There is overwhelming evidence that Shakespeare intended to create lines of ten, or sometimes (just as intentionally) eleven, syllables, in nearly every one of his hundreds of sonnet lines (and where he fell short, as in the eight-syllable lines of sonnet 145, he was clearly counting too).[49] This evidence confirms the judgment of Hemminge and Condell, in the preface to the First Folio, that Shakespeare's works are "absolute in their numbers."

Yet it also appears that Shakespeare had some notion of writing verses in feet. His plays are filled with references, especially punning references, to poetic feet that run, stop, skip, limp, or stumble: "This is the very false gallop of verses; why do you infect yourself with them?" (*As You Like It*, 3.2.113–14); "[T]he feet were lame, and could not bear themselves without the verse, and therefore stood lamely in the verse" (*As You Like It*, 3.2.169–71); "a halting sonnet" (*Much Ado about Nothing*, 5.4.87). Yet these recurring allusions to the "movement" of feet, which seem, as Puttenham would have it, to refer to "quantity" or the time it takes to pronounce syllables, refer instead, once again, to syllable count: "I heard them all, and more too, for some of them had more feet than the verses would bear" (*As You Like It*, 3.2.164–66). A "lame" verse, in other words, probably refers to a line that has more or less than ten (or sometimes eleven) syllables, not to one that lapses from iambic or any other foot-based meter. The line, mentioned earlier, that Jacques labels as blank verse—"Good day and happiness, dear Rosalind!"—is only "licentiously" iambic in its rhythms. Yet because it consists of exactly ten syllables, Jacques seems to mean unrhymed decasyllabics, just as Nashe refers to "blank verse" interchangeably with "drumming decallabon[s]."

Shakespeare's own references to poetry do not suggest an awareness of writing iambic verses or in lines organized by classical feet. The best evidence we have for characterizing Shakespeare's meters as iambic derives not from what he says but from what we make of his practice: count-

---

[49] Paul Ramsay, "The Syllables of Shakespeare's Sonnets," in *New Essays on Shakespeare's Sonnets*, ed. Hilton Landry (New York: AMS Press, 1976), 193–215.

less lines in the *Sonnets* scan that way (again, as long as we substitute stress for duration as the basis for defining poetic feet):

When I consider everything that grows
[Holds in perfection but a little moment;]
That this huge stage presenteth nought but shows
Whereon the stars in secret influence comment.
(sonnet 15.1–4)

When most I wink then do mine eyes best see,
For all the day they view things unrespected,
But when I sleep, in dreams they look on thee,
And darkly bright, are bright in dark directed.
(sonnet 43.1–4)

In a systematic alternation that recurs elsewhere in the *Sonnets*, lines 1 and 3 of these poems are decasyllabic, while lines 2 and 4 are hendeca-syllabic; Shakespeare without exception keeps constant the number of syllables occurring in rhyming lines (of either ten or eleven syllables each), suggesting, once again, a conscious concern for these numbers. But what about the "regular" rhythm of these lines? According to foot-based methods of scansion (which presume, once again, that Shake-speare transferred ideas about quantity to matters of stress) variations occur by means of deliberate substitutions of poetic feet. Most metricists claim, for example, that Shakespeare substitutes a trochee for an iamb in the first foot of the following lines:

Pity the world, or else this glutton be   (1.13)

Look in thy heart, and tell the face thou viewest   (3.1)

Bearing thy heart, which I will keep so chary   (22.11)

Duty so great, which wit so poor as mine   (26.5)

Nothing, sweet boy, but yet like prayers divine   (108.5)

There's no doubt that certain metrical alterations, like this one, occur frequently enough to suggest a deliberate pattern of variation. Indeed, given its explanatory power, there is no way positively to dispense with feet in our approach to Shakespeare's meters.

Yet abiding by feet undoubtedly makes the metrical analysis of certain lines more difficult, if not impossible:

Let me not to the marriage of true minds
Admit impediments.
(116.1–2)

These lines (once again, of strictly ten syllables each) cannot be scanned as a sequence of feet, at least not without seriously compromising the sense (I leave it to readers to try as they might). It seems unlikely that Shakespeare created the random, proselike rhythms of these lines by systematically substituting trochees, or dactyls, for iambs. Nor does it doesn't help the case for poetic feet that Shakespeare's prose sometimes scans as iambic, just as his verse does, in passages such as the following:

Why then 'tis none to you; for there is nothing either good or bad, but thinking makes it so. To me it is a prison. (*Hamlet*, 2.2.249–50)

The beauty of the world; the paragon of animals (*Hamlet*, 2.2.307)

To tell you where he lodges, is to tell you where I lie (*Othello*, 3.4.8–9)

To bed, to bed, there's knocking at the gate. (*Macbeth*, 5.2.66–67)

Is man no more than this? . . . Thou ows't the worm no silk, the beast no hide, the sheep no wool, the cat no perfume. . . . Thou art the thing itself. (*Lear*, 3.4.102–6)

As Wright concedes, "from a formal point of view, the most notable aspect of Shakespeare's prose is that it is hard to distinguish from verse."[50] There are numerous passages in the plays which may be scanned as either verse or prose, as the history of editorial practice on the plays bears witness. Is it possible, then, that Shakespeare was measuring feet in his prose? It is hard to imagine: Shakespeare's "non-iambic" prose, like his variable verse, would have to arise as the result of foot substitutions, of methodical variations of trochees and dactyls from an iambic template. If it is impractical to proceed as if Shakespeare measured his "iambic" prose into feet, it is possible, at least, that he composed his "iambic" verse without recourse to them as well. I am reluctant to make the strong claim that there are no feet in Shakespeare's poems, or in any poems, as

---

[50] Wright, *Shakespeare's Metrical Art*, 110.

Attridge so persuasively does. But the ontological status of certain meters—whether there "are" such things as feet, for example—is not really
the question here. The problem, instead, is the extent to which modern
metrics mismeasures Shakespeare's verse insofar as it counts his lines in
ways the poet never conceived.

The regularity of Shakespeare's stress patterns in the *Sonnets*—the
fact that we are able systematically to align his verses with our favored native measure—makes it impossible to refute to the iambic pentameter
hypothesis. It is, however, still only a hypothesis: although there is considerable evidence that Shakespeare counted syllables in his sonnet lines,
there is no comparable evidence that he deliberately counted feet. We
must allow for the possibility that poetic feet may be standing in the way
of our fully understanding Shakespeare's meters, at least as he might
have imagined them himself. In what follows I present some alternatives
to foot scansion as an explanation of Shakespeare's metrical effects. To
judge Shakespeare's metrics appropriately, we must try to set aside our
own instruments and take up Renaissance terms of poetic measure, especially, once again, the importance of proportion or relational "parts"
in the making of metrical value.

What animates the study of Shakespeare's rhythms, ultimately, is the
sense that his metrics "mean" something, that they represent or perform
aspects of the poetic personae who use them. It is rarely implied that his
characters deliberately manipulate their own meters (although this
might make for a promising new point of departure for analysis); rather,
Shakespeare is presumed to measure his characters, expose their natures, via his metrics. The attempt to discover a semantics of metrical
form is a central component of rhetorical analysis, but one that remains,
perhaps, the most uncertain—in part because the relationship of "measure" to meaning is never a stable one. Like his figuration, Shakespeare's
metrics is fundamentally a contingent, comparative art.

Wright calls Shakespeare's metrical practice in the *Sonnets* "an art of
small differences," a characterization that is fair enough from any formal
standpoint, and shows how small differences can produce powerful effects. He notes, for example, the "emotional disturbances" created by
variations of the meter of sonnet 29:

When in disgrace with fortune and men's eyes,
I all alone beweep my outcast state,
And trouble deaf heav'n with my bootless cries,

And look upon myself and curse my fate.
(1–4)

Wright explains that in this poem "the meter tends to depart from the normal pattern in the odd lines and to return in the even lines":

> In the first quatrain [above], line 1 incorporates an initial trochee, a pyrrhic, and a spondee, but line 2 returns us to normal, varying the rhythm only with a final spondee; line 3 goes further afield than line 1, rising to a height in the monosyllabic "heaven."... The return of line 4 to a very regular rhythm seems, in contrast and in context, expressive of the speaker's despair.[51]

Later in the sonnet, as Wright explains it, Shakespeare once again departs from a strict iambic pentameter only to return to it:

> Yet in these thoughts myself almost despising,
> Haply I think on thee, and then my state,
> (Like to the lark at break of day arising
> From sullen earth) sings hymns at heaven's gate.
> (9–12)

After several lines that include trochaic reversals ("almost"; "Haply"), the "resumption of regular meter in the last four feet of line 10 no longer carries the tone of melancholy saw earlier in even lines; rather, it seems on the verge of something imminent and admirable." Wright makes explicit his view that the same meter may represent opposing emotional states: "What is remarkable here is the poet's power to make the same metrical effects convey quite opposite feelings as the poem's outlook alters."[52] This, however, gives the game of "interpretive metrics" away: there is no intrinsic "meaning" in a measure. Its interpretation depends on the *relation* of one metrical pattern to another *in a given context*.

If deriving meaning from Shakespeare's numbers in the *Sonnets* depends on the "small [metrical] differences" that distinguish one line from the next, interpreting the meters of his plays is far more difficult,

---

[51] Ibid., 75, 80.
[52] Ibid., 81.

given the range and variety of "regular" and "irregular" lines that occur. The line of Shakespeare's late plays has been described as an "immensely flexible blank verse iambic pentameter"—so immensely flexible, in fact, that "in Shakespeare's later plays . . . we rarely encounter a passage of more than a few lines that does not break the pattern in some way." Therefore "any account of Shakespeare's iambic pentameter has to cope with the great variety, almost the miscellaneousness, of lines that appear in and around the central verse type."[53] Not surprisingly, there is no way to "fix" the psychological or emotional "nature" of the "central verse type," nor any of its variations. For Hardison, Hamlet's "To be or not to be" soliloquy is "intentionally lacking in melody" because it asks the audience to share in a "reasoning" process.[54] Yet Woods hears same lack of music in *The Tempest* as a sign of malevolence: "[T]he villainy of Antonio and Sebastian becomes evident not only in their cynicism and their apparent inability to see the beauties the others see, but also in their relentlessly prosaic speech rhythms."[55] Strict iambic pentameter can apparently signal judiciousness, as in Gonzalo's speeches in *The Tempest*; at other times it is taken as evidence of distress and insecurity, as when Claudius and Gertrude, in the face of Ophelia's mad rambling, work "to keep up iambic appearances."[56] Hardison reminds us that the metrical effects of a given speech will depend in part on theatrical delivery: "Macbeth's 'Tomorrow and tomorrow and tomorrow' speech can be richly and darkly musical, or it can be fragmented and devoid of melody—an objectification of the emptiness of despair."[57]

Hardison, however, stands with the consensus on the matter of why Shakespeare's late meters vary so much: the playwright was developing a more "natural" line. Hardison uses *Hamlet* to illustrate the difference between Shakespeare's early and late metrics. We can hear that difference, according to Hardison, in the contrast between the stilted speeches of the "players" within the play and the more natural speeches of Hamlet himself.[58] The Player's speech, he suggests, represents an old-fashioned measure. Yet it includes numerous lines that scan as regular lines of iambic pentameter:

---

[53] Ibid., 101, 103.
[54] Hardison, *Prosody and Purpose*, 255.
[55] Woods, *Natural Emphasis*, 249.
[56] Wright, *Shakespeare's Metrical Art*, 104.
[57] Hardison, *Prosody and Purpose*, 255.
[58] Ibid., 254.

> The rugged Pyrrhus, he whose sable arms (2.2.452)
>
> Hath now this dread and black complexion smear'd (455)
>
> With heraldy more dismal: head to foot (456)
>
> With blood of fathers, mothers, daughters, sons (458)
>
> With eyes like carbuncles, the hellish Pyrrhus (463)

Many others begin with a stressed syllable (i.e., in traditional scansion, with a trochaic substitution):

> Black as his purpose, did the night resemble (453)
>
> Bak'd and impasted with the parching streets (459)

Others break further from the pattern:

> Now is he total gules, horridly trick'd (457)
>
> To their lord's murther. Roasted in wrath and fire (461)

And some lines are short:

> Did nothing (482)
>
> Now falls on Priam (492)

Hamlet's soliloquy, immediately following the player's speech, shows the same regularities and includes comparable variations. "Iambic" lines include

> O, what a rogue and peasant slave am I! (550)
>
> That from her working all the visage wann'd (554)
>
> What's Hecuba to him, or he to Hecuba,
> That he should weep for her? What would he do (559–60)
>
> And can say nothing; no, not for a king (569)

Many of his lines begin with a stressed syllable:

> Tears in his eyes, distraction in his aspect (555)
>
> Who calls me villain, breaks my pate across (572)
>
> Plucks off my beard and blows it in my face (573)

And there are several short lines:

> For Hecuba!   (558)

> About, my brains! Hum—I have heard   (588)

The rhetorical difference between Hamlet's speech and the speech of the player has little to do with meter. If the player's speech seems artificial, it is because of its inflated diction, including phrases such as "o-er-siz'd with coagulate gore" (462); "anon the dreadful thunder / Doth rend the region" (486–87); "with bisson rheum" (506); and, most famously, "the mobled queen" (502). Metrically speaking, they are comparable examples of Shakespeare's mature art.

Indeed, the perception that Shakespeare's meters become more "natural" in his later plays may have less to do with his verses and more to do with the playwright's increasing use of prose. As is well known, Shakespeare successively loosened the social tie that made prose appropriate only for lower-class characters in drama. Wright maintains that Shakespeare's late prose "remains, by and large, a resource for the most dignified characters to use only at moments of suspended intensity."[59] Yet "fixing" it this way does not help us understand why prose is also so often the language of familiar companionship (as in the taverns of the *Henry IV* plays) or of madness (in *Hamlet, Macbeth,* and *Lear*). Prose doesn't "mean" one thing any more than verse does; rather, they mean in increasingly complex relations to each other, as is especially clear when characters shift from one metrical mode to the other.

The most famous of such shifts may well be Prince Hal's sudden turn to verse in 1.2 after a long scene of bantering in prose with Falstaff in the tavern; in his first soliloquy, he not only switches to metered speech but makes explicit some corresponding shift in identity (195–217). Hal's shift from prose to verse may signal a revelation of his "true," socially superior self. If that's so, Shakespeare conforms here to early, social conventions regarding dramatic meters. Crucially, however, we cannot assume that these conventions hold beyond this particular dramatic context, especially in the later plays. Caliban in *The Tempest*, after all, speaks in prose in a scene with Stephano and Trinculo, shifting to verse only in an aside (2.2.116–18). Whatever this variation might mean, it is

---

[59] Wright, *Shakespeare's Metrical Art*, 109.

not a sign of an elevated social status. In the opening scene of *Coriolanus*, the citizens use prose in their conversation with the patrician Menenius, while he responds to them in verse, but they switch to verse when they question him on his famous parable of the "body politic." In 1.3, Volumnia and Virgilia speak to one another in prose, then shift to verse when a Gentlewoman enters the scene; after the Gentlewoman exits, they continue in verse for a few lines before returning to prose. Once again, whatever these alternations indicate about the characters who make them, they are not predetermined; their meanings are generated at the most local level of the text. By proportioning prose and verse to one another in varying relations, Shakespeare suggests once more that their "meaning" inheres, if it inheres in anything, only in those relations themselves.

Besides his increasing use of prose, or unmetered language, in his late plays, Shakespeare's most frequent deviation from the decasyllabic line is the short line. About one in every twenty-four lines in Shakespeare's plays is short (i.e., of fewer than ten syllables in length), whether it is written in verse or prose.[60] Wright explains the effect this way: "[T]he uncompleted short line, even at the end of an orderly speech, suggests a momentary breakdown of the system . . . the effect can be almost that of a runaway language, still iambic but alarmingly unstrung."[61] I would like to conclude this chapter by reconsidering the effects of such lines and what they may tell us about Shakespeare's poetic "measures" generally. Here are some examples:[62]

> So sweet was ne'er so fatal. I must weep,
> But they are cruel tears. This sorrow's heavenly,

---

[60] See E. K. Chambers, *William Shakespeare: A Study of Facts and Problems* (Oxford: Clarendon Press, 1930).

[61] Wright, *Shakespeare's Metrical Art*, 141.

[62] I have excluded among my examples "short" lines that are metrically "completed" by a line, spoken by another character, that follows it. For example,

1 Gent. Whither away so fast?
2. Gent. O, God save ye!
(*Henry VIII*, 2.1.1)

Cassius. You bear too stubborn and too strange a hand
Over your friend that loves you.
Brutus. Cassius.
(*Julius Caesar*, 1.2.35–36)

It strikes where it doth love. She wakes.
(eight syllables; *Othello*, 5.1.20–22)

Expose thyself to feel what wretches feel,
That thou mayst shake the superflux to them
And show the heavens more just.
(six or seven syllables; *King Lear*, 3.4.34–36)

Nay, nay, Octavia, not only that—
That were excusable, that, and thousands more
Of semblable import—but he hath wag'd
New wars 'gainst Pompey, made his will, and read it
To public ear.
(four syllables; *Antony and Cleopatra*, 3.4.1–5)

Life's but a walking shadow, a poor player,
That struts and frets his hour upon the stage,
And then is heard no more. It is a tale
Told by an idiot, full of sound and fury,
Signifying nothing.
(six syllables; *Macbeth*, 5.5.24–28)

Rather than a "breakdown of the system," these lines may expose a new kind of system altogether, an experiment that Shakespeare, renowned as he is for his "natural" measures, is no doubt the last one we would expect to attempt it: quantitative metrics.

I don't mean this in the strictest, classical sense: Shakespeare did not treat English syllables as if they were, by their *nature*, of different lengths, as measured by time. His short lines, however, suggest an effort to manipulate the sense of a word's length or the length of the line as a whole. For Shakespeare, quite possibly, short lines are created so as to "weigh" as much as the long lines to which they compare. Macbeth's final line about life "signifying nothing," for example, may last just as long, in emphasis, as the full ten or eleven syllables that precede it. The line, in other words, is not so much short as "stretched"; proportionally it is "equivalent" to a "regular" decasyllabic or hendecasyllabic verse. There are two ways in which Shakespeare may establish these equalities. Short lines may contain unrealized (i.e., silent) beats that make up a full line; alternatively, Shakespeare may intend us to "hear" syllables of the short line as "stretched," that is, as taking more time than the syllables of regular lines. Either way, short lines would thus not be "short" at all: by proportioning Macbeth's "signifying nothing" to the measure of a longer line, Shakespeare creates what can only be described as a *quantitative* effect.

There are many short lines that seem to relate this way to full-length lines and that are designed to seem equivalent in weight to them. A good example is from *Lear*:

> *Blow, winds, and crack your cheeks! rage, blow!*
> You cataracts and hurricanoes, spout
> Till you have drench'd our steeples, [drown'd] the cocks!
> You sulph'rous and thought-executing fires,
> Vaunt-couriers of oak-cleaving thunderbolts,
> Singe my white head! And thou, all-shaking thunder,
> Strike flat the thick rotundity o'th'world!
> (3.2.1–7)

For all his fury, Lear manages to speak in lines made almost entirely of ten or eleven syllables each. Yet the first line of this speech is only eight syllables long. Is it certain, as others have suggested, that these lines are incomplete, that they represent a metrical "lack"? Given the poet's metrical terms, it seems more likely that Shakespeare intended Lear's impassioned imperatives (his eight, angry monosyllables) to "add up," in a deliberate counterpoise, to a full line. To take another example,

> Art thou not, fatal vision, sensible
> To feeling as to sight? or art thou but
> A dagger of the mind, a false creation,
> Proceeding from the heat-oppressed brain?
> I see thee yet, in form as palpable
> *As this which now I draw.*
> Thou marshal'st me the way that I was going,
> And such an instrument I was to use.
> (*Macbeth*, 2.1.36–43)

Again, a speech consistently made of lines of ten or eleven syllables includes one short line, "As this which now I draw," only six syllables long. But isn't it possible that Shakespeare "stretched" these syllables, heavily, ominously, over the duration of the line, or, alternatively, that he allowed for several syllables of silence in which Macbeth draws out his "real" dag-

ger? As a final example, here is Hamlet's speech in response to the
Player's performance in act 2, scene 2:

> Is it not monstrous that this player here,
> But in a fiction, in a dream of passion,
> Could force his soul so to his own conceit,
> That from her working all the visage wann'd,
> Tears in his eyes, distraction in his aspect,
> A broken voice, an' his whole function suiting
> With forms to his conceit? And all for nothing,
> *For Hecuba!*
> What's Hecuba to him, or he to [Hecuba],
> That he should weep for her? What would he do
> Had he the motive and [the cue] for passion
> That I have? He would drown the stage with tears,
> And cleave the general ear with horrid speech,
> Make mad the guilty, and appall the free.
> (2.2.551–64)

"For Hecuba" is conventionally scanned as a (very) short line, yet it is
surely "lengthened" by the time Shakespeare leaves for Hamlet's indignation and despair. In his late plays, Shakespeare may well have found a
way to create the very illusion that Daniel later decried—"those imagined quantities of syllables."

There is simply no definitive evidence that Shakespeare himself heard in
his verses what so many have heard since: iambic pentameter. It would be
surprising if, for example, Shakespeare consciously substituted trochees
for iambs; he does not seem to know this terminology, in reference to
stress rather than quantity, even if it was conceptually in the making. We
can, however, be sure of two things regarding Shakespeare's metrics: (1)
that he was counting syllables, as vernacular metrics throughout the
Renaissance prescribed; and (2), that he inherited a system of poetic
measurement that privileged proportion, above all, as the highest value
of verse. There is no doubt that Shakespeare's line becomes more natural (if by "natural" we mean more like speech) as it becomes increasingly
hard to distinguish from (unmetered) prose. Yet to the extent that the
playwright was working to develop a new meter, his measures become
more complex, more "artificial," as Renaissance poets would judge them.
Over the course of his metrical career, the playwright abandoned his "art

of small differences" in favor of more exaggerated disproportions of sound (as when, for example, his characters shift between verse and prose). And he devised new "ratios of [metrical] equality": as his use of so-called short lines, especially, suggests, he varied his meter to produce *effects* of equivalence analogous to the proportioning of weights of syllables in classical verse.

Shakespeare, no less than Sidney or Spenser, attempted a kind of "quantitative" verse. He developed English numbers (in Renaissance terms, decasyllabic verse) in ways that went beyond mere counting. His *ars metrica* measures words "geometrically" rather than "arithmetically," as dynamic and relational rather than sequenced and set by degree. However natural his rhythms seem, however "native" to the language, he chose to adapt a foreign system of measurement to help create them. As Samuel Daniel acknowledged, metrics is always, must always be, "foreign" to speech, a mismeasure of "real" language, just as "figures" of speech reimagine (human) relations. As Shakespeare credits them, however, the rules of poetry create the impression, at least, that we can enumerate our lives, reckon our loves—to the last syllable of recorded time.

POUNDS OF FLESH

Race Relations in the Venice Plays

"A LITTLE MORE THAN KIN, and less than kind" (1.2.65)—so Hamlet defines his new relation to his uncle Claudius, now also his father since having precipitously married the prince's widowed mother, Gertrude. As his notorious quip is usually glossed, Hamlet responds to Claudius's double salutation, "now my cousin Hamlet, and my son," by identifying himself as "closer than a nephew . . . yet more distant than a son, too (and not well disposed to [you])"[1] Hamlet's pun on "kin" and "kind," however, suggests a more pointed, productive coincidence of terms. More than just affined by blood ("a little *more* than *kin*") yet less than equivalent in being ("*less* than *kind*"), the two men's filiation is based on changing degrees—"more" or "less"—of relatedness. Hamlet's word "kind" literally enacts an addition to the word "kin," and yet, crucially, there exists no term between kin and kind that can precisely or conclusively denominate their new relationship.

Hamlet's famous pun unwittingly performs the riddle of Renaissance notions of race: How can a man be more than kin to another and yet less than of his kind? What does it mean to posit differing degrees or increments of human nature, especially, if those increments may lessen or accrue? This chapter will explore early modern human measurements based on the body, evaluations of human value determined by pounds of human flesh. What's at stake in part is the notion, already current in the sixteenth century, of a human "equality by nature," just *how* equal men and women may be judged to be, pound for pound of human flesh. In

---

[1] William Shakespeare, *The Riverside Shakespeare*, 2nd ed., ed. Blakemore Evans (Boston: Houghton Mifflin, 1997), n. 65.

my focus on human relations, once again, as the determinant of human value, I will also suggest something about Shakespeare's understanding of race: rather than a permanent material condition, a discrete, fixed, or invariable nature, race for Shakespeare names a relation that arises *between* people; it is a kinship, but one that is generated among potentially shifting points of family resemblance.[2]

We are often cautioned against anachronism in studies of race and ethnicity in the early modern world, and against the danger of imposing modern categories and setting modern boundaries around the diversity of Renaissance human-kinds.[3] Although negotiating the differences among notions of race, religion, and nationality in the Renaissance may be difficult, language offers an intervention. The early modern English discourse of human "kinds" includes terms that designate particular groups, for example, Moor, Indian, or Turk. From a modern perspective, membership in such groups may be fluid and often transgresses our own racial and national lines.[4] This discourse, however, also includes words that pertain to race, religion, and nationality more generally. The semantic field of Renaissance words relating to the peoples of the world encompasses the following terms: *people, nation, stock, stem, race, line, line-*

[2] My notion of "race relations" revises an older, well-established critical formulation regarding the relationship between the Self and the Other, in which the Other is a projection of the Self and arises in opposed relation to it. This older construct privileges the Self as the point of origin for the production of otherness; my emphasis on *relationality*, in contrast, suggests the mutual interdependence of two or more persons whose identities are compared. Whereas the terms Self and Other posit their relation as one of absolute difference, "race relations" allows for any and all degrees of "equality" or "inequality" among its terms. Finally, although Self and Other are abstractions (because these terms name a structural relation of absolute difference, it is not essential to specify the kind of "otherness" imagined in particular cases), "race relations," which names shifting values, depends on highly specific and unstable cultural comparisons.

[3] For a historical survey of the evolution of ideas about race, including a discussion of the nature of racialized thinking in the early modern world, see Ivan Hannaford, *Race: The History of an Idea in the West* (Baltimore: Johns Hopkins University Press, 1996). For recent warnings against anachronism in the study of Renaissance race, see Michael Neill, "'Mulattos,' 'Blacks,' and 'Indian Moors,': *Othello* and Early Modern Constructions of Human Difference," *Shakespeare Quarterly* 49.2 (1998): 361–74; and Ania Loomba, *Shakespeare, Race, and Colonialism* (Oxford: Oxford University Press, 2002).

[4] To take just one example, many early modern writers typically conflate or confuse Moors with Indians. The Spanish Moor Eleazar, in *Lust's Dominion or The Lascivious Queen*, ed. Khalid Bekkaoui (Fez, 1999), is called an "Indian slave" and swears "[b]y all [his] Indian gods" (3.4, p. 47; 4.3, p. 110). All further citations of *Lust's Dominion* refer to this edition.

age, descent, breed, brood, kind, kin, tribe. The majority of these words share a common lexical feature—the sememe of *nature,* a relationship of birth or blood.[5] What emerges here is a notion of race based securely in familiality, but within which specific groups are potentially reconfigured or reconstituted.

In Venice, amid "the concourse of strange and forraine people, yea of the farthest and remotest nations,"[6] Shakespeare interrogates two of the tribes of Renaissance Man, Jews and Muslims. There is nothing coincidental about the fact that Shakespeare considered Jews and Muslims in succession in Venice, first in *The Merchant of Venice* and later in *Othello, the Moor of Venice.* There is much evidence, both historical and literary, that Renaissance Christians saw these "nations" as closely related ones. But just how closely related? What one scholar has called the medieval and early modern "Jewish-Muslim symbiosis" is essential to understanding Shakespeare's interest in the "scattered nations" of Jews and the Moors, who settled in early modern Venice and appear, a decade apart, in Shakespeare's only two plays set in that city.[7] I turn now to some of the human measurements taken in Shakespeare's Venice plays, especially the "natural" kinships imagined among Christians, Jews, and Muslims. For Shakespeare, these often prove, in relation to one another, a little more than kin, and less than kind.

SHYLOCK: This is kind I offer.
BASSANIO: This [Shylock's offer] were kindness.

---

[5] See Ania Loomba, "The Vocabularies of Race," in Loomba, *Shakespeare, Race and Colonialism,* 22–44. Loomba similarly emphasizes the semantic ties between race and "lineage" in early modern English. Mary Floyd Wilson, *English Ethnicity and Race in Early Modern Drama* (Cambridge: Cambridge University Press, 2003), coins the term "geohumoralism" to describe her theory that Renaissance writers understood race to be based in a nexus of region, climate, and bodily humors. Janet Adelman, "'Her Father's Blood': Race, Conversion, and Nation in the Merchant of Venice," *Representations* 81 (Winter 2003): 4–30, also emphasizes the importance of "blood" and "lineage" as the basis of what she describes as Shakespeare's "protoracialism."
[6] Gaspar Contareno, *The Commonwealth and Gouernment of Venice* (London, 1599), 3.
[7] The phrase "Jewish-Muslim symbiosis" is Avigdor Levy's, as quoted in Daniel Vitkus, ed., *Three Turk Plays from Early Modern England: "Selimus," "A Christian Turned Turk," and "The Renegado"* (New York: Columbia University Press, 2000), 37. See also Loomba, *Shakespeare, Race and Colonialism,* 146–48, on the perceived alliance between Jews and Muslims in the Renaissance; and Adelman's review of recent scholarship on the early modern identification of Jews with "blackness," in "Her Father's Blood," 27n27.

SHYLOCK: This kindness will I show.
(1.3.141–3)

ANTONIO: [I'll] say there is much kindness in the Jew.
(153)

ANTONIO: The Hebrew will turn Christian, he grows kind.
(178)

There is a recurring pun on the word "kind" in act 1, scene 3, of *The Merchant of Venice*. The two meanings in play here, "type" or "nature," and "compassion," are inseparable in the racial semantics operating among the Christian characters of the play. To them, for Shylock to show "kindness" is for him to "turn Christian," that is, not only to act in a Christian manner (here, out of mercy rather than vengefulness) but to become Christian or perhaps even to become human. (In a play, and in a culture, where Jews are considered "damn'd, inexecrable dog[s]" and "inhuman wretch[es]," it amounts to the same thing [4.2.128; 4.1.4].) Shylock, however, is only pretending to be kind, playing at Christian compassion, dissembling their "humanity," when he proposes the bond of human flesh. He is playing at the converso he will later be compelled to become. And crucially, this moment of deception, of feigning "kindness," is precisely the moment that Shylock becomes most dangerous. It is one thing for Shylock to be a Jew and clearly apprehended as such. Far more threatening, as Antonio and Bassanio are soon to learn, is for Shylock to be a dissembling "New Christian" who combines "the public persona of a Christian with the secret beliefs and rituals of a Jew."[8] Although Shylock is not subject to the Inquisition of Venice, his trial makes him subject to a Christian court that ultimately will seek to enforce "kindness" if it cannot elicit it unstrained. It seems likely, indeed, that Shakespeare inflected Shylock's trial in act 4, which finally turns on his "alien" nature,[9] with intimations of the Inquisitions of early modern Spain and Italy.

Yet the natural "kindness" of Christian and Jew is exactly what Shylock affirms in his infamous rationale for vengeance. For convenience I cite the controversial passage here:

[8] Brian Pullan, *The Jews of Europe and the Inquisition of Venice, 1550–1670* (Totowa, NJ: Barnes and Noble, 1983), 123.
[9] Portia, disguised as the learned judge Balthazar, notoriously cites a special law, directed only against "aliens," that sentences Shylock to death for attempting the life of a Venetian (4.1.349).

Hath not a Jew eyes? Hath not a Jew hands, organs, dimensions,
senses, affections, passions; fed with the same food, hurt with the
same weapons, subject to the same diseases, heal'd by the same
means . . . as a Christian is? If you prick us, do we not bleed? If you
tickle us, do we not laugh? If you poison us, do we not die? And if
you wrong us, shall we not revenge? If we are like you in the rest,
we will resemble you in that. . . . The villainy you teach me, I will
execute, and it shall go hard but I will better the instruction.
(3.1.59–73)

Once seen as a moving plea for compassion on the grounds of the common humanity of Christian and Jew, Shylock's speech is now more often
condemned for an apparently cynical or even treacherous turn of logic
at its conclusion. Having spoken of shared pleasures and pains, Shylock
proceeds to defend his intention to kill Antonio on the grounds that Jews
resemble Christians in their desire and determination to exchange
wrong for wrong. Some have censured Shylock's reasoning here, citing
him for sophistry, while others see the conclusion of his speech as proof,
finally, of Shylock's moral depravity, indeed, of the inhumanity of the Jew.
Thus Shylock's "assertion of similarity, far from humanizing him, demands to be seen as a threat of vengeful imitation based on the illegiti-
mate code of 'an eye for an eye.' "[10] According to this view, Shylock's reminder that he too has eyes and feels pain has nothing at all to do with
"fellow feeling"; rather, it reveals the sort of "imitative exchange" and violent doubling that structures the *lex talionis* or Old Testament code of
retaliation.

There is, however, no such turn in Shylock's famous speech. From beginning to end, it is grounded in commonplace sixteenth-century notions of a human "equality by nature," that is, an equality based on Renaissance measures of human bodies and, moreover, on concurrent ideas
of revenge as a natural, bodily desire to maintain or re-create that equality. Shylock's rhetorical question "Hath not a Jew eyes?" is a natural and
proper preamble to his insistence on "an eye for an eye" justice because
the desire for revenge, however morally unacceptable, was understood to
be of a kind with other affections and passions of the human body.

---

[10] Jonathan Gil Harris, *Foreign Bodies and the Body Politic: Discourses of Social Pathology in
Early Modern England* (Cambridge: Cambridge University Press, 1998), 100.

For Renaissance English writers, the original exponents of the idea of "natural equality"—the notion that people should be deemed equal because of shared physical qualities or capacities present at birth—were the Roman Stoics. In response to an Aristotelian social and political ethic grounded in the notion of the inequalities of human nature, the Stoics argued for a "natural law" of universal reason, or at least a universal capacity for reason, innately given, which levels social differences between master and slave, man and man. Cicero and Seneca, perhaps the most influential advocates for the idea that "reason is common to all; men differ indeed in learning, but are equal in the capacity for learning," concluded that "[t]here is no resemblance in nature so great as that between man and man, there is no equality so complete."[11] Seneca, perhaps to reconcile his belief in human equality with the realities of social and political life in Rome, concluded that "natural equality" only prevailed in an earlier age, a golden age in which "nature" held sway. Roman law ultimately codified the notion that "all men are equal" according to natural law.[12]

Early Christian writers reaffirmed the notion of human equality. As Augustine wrote, "[W]hoever is anywhere born a man, that is, a rational mortal animal, no matter what unusual appearance he presents in color, movement, sound, nor how particular he is in some power, part or quality of his nature, no Christian can doubt that he springs from one protoplast."[13] Yet in Christian thought, natural equality is a function not so much of birth as it is of rebirth through, and as, the "one body" of Jesus. Saint Paul pronounced, "There is nether Iewe nor Grecian: there is nether bonde nor fre: there is nether male and female: for ye all are one in Christ Iesus."[14] Paul's assertion that all men are equal insofar as they share a common descent from God, and a shared being in the body of Christ, may have some lineal relationship to classical notions of equality by nature, but the revival of the idea of "natural law" and the effort to assimilate it to Christian theology begin unequivocally with Thomas

---

[11] See R. W. Carlyle and A. J. Carlyle, *A History of Medieval Political Theory in the West*, 6 vols. (Edinburgh: William Blackwood, 1903–36), 1:7, 8.

[12] As recorded in the Digest, "According to natural law, all men are equal," as quoted in Paul E. Sigmund, "Hierarchy, Equality, and Consent in Medieval Christian Thought," in *Equality*, ed. J. Roland Pennock and John W. Chapman, 134–53 (New York: Atherton Press, 1967), 138.

[13] Quoted in Hannaford, *Race*, 96.

[14] Gal. 3:28; cf. 1 Cor. 12:13 and 1 Col. 2:11.

Aquinas in his *Summa theologiae*.[15] By the thirteenth century, the idea of a universal human nature, located in the human body, was a commonplace of Christian thought. Gregory the Great's dictum, "by nature all of us are equal" [omnes namque nature aequales sumus], was widely cited throughout the Middle Ages.[16] The author of the *Roman de la rose*, for example, presents it as an incontrovertible truth:

> Naked and impotent are all,
> High-born and peasant, great and small.
> That human nature is throughout
> The whole world equal, none can doubt.[17]

Yet however common the refrain, it must be noted that those who repeated it did not necessarily agree about the consequences of natural equality for social life. Early modern writers who begin with the premise of human equality by nature may otherwise be at odds, politically or ideologically. Montaigne might well have spoken Machiavelli's words, "All men, having had the same beginning, are equally ancient and have been made in one mode. Strip us all naked, you will see us all alike."[18] Richard Hooker might have concurred, despite greater doctrinal differences, with Martin Luther when he asked, "Whence comes this great distinction between those who are equally Christians? Only from human laws and inventions!"[19]

By the Renaissance, however, "equality" not only names a natural state of being but a natural feeling about being in relationships with others. From Galen and his Renaissance heirs, from Cicero, and from Seneca, Renaissance English writers understood that all emotions are corporeal in nature: "You will have no doubt, I am sure, that emotions

[15] R. S. White, *Natural Law in English Renaissance Literature* (Cambridge: Cambridge University Press, 1996), 29.
[16] As Ulpian declared, "quod ad jus naturale attinet, omnes hominess aequales sunt," as quoted in Carlyle and Carlyle, *History of Medieval Political Theory*, 1:114.
[17] Quoted in Sigmund, "Hierarchy, Equality, and Consent," 140.
[18] Niccolò Machiavelli, *Florentine Histories*, trans. Laura F. Banfield and Harvey C. Mansfield Jr. (Princeton: Princeton University Press, 1988), 122.
[19] Quoted in Sanford A. Lakoff, "Christianity and Equality," in Pennock and Chapman, *Equality*, 115–33, quote at 129.

are bodily things."[20] Emotions or "griefs," according to the Stoics, are pains or diseases of the body, or the soul understood as a part of the body. It is well known that disease is considered endemic to the body in the Galenic system; less often noted is the way that some griefs arise as sympathetic or antipathetic responses to the pains, and pleasures, of other bodies. Envy, for example, "is the worst grief. . . . I call it envy whenever someone is grieved over the success of others."[21] Mercy, on the other hand, "is a greefe whiche a man takes for the paines, miseries, or aduersities of another."[22] Vengefulness too is a grief felt in relation to the body of another, as Pierre de la Primaudye explains in *The French Academie*: "When the heart is wounded with griefe by one, it desireth to returne the like to him that hath hurt it, and to rebite him of whome it is bitten. Every offense therefore that ingendereth hatred, anger, enuy or indignation bringeth with it a desire of reuenge, which is to render euill for euill, and to requite griefe receiued with the like againe."[23] The human heart, being "wounded with griefe by one," desires to inflict "the like" wound in return; revenge may be morally wrong (as de la Primaudye finally asserts), but its ultimate source is not the "old laws" of ancient Greece, or Rome, or Israel, but the enduring laws of the human heart. That is why the ghost of Old Hamlet incites his son to revenge out of natural desire ("If thou hast nature in thee, bear it not" [1.5.81]), and why Laertes distinguishes the matter of honor from that of "nature" as motivating sources for revenge, citing the latter as primary: "I am satisfied in nature, / Whose motive in this case should stir me most / To my revenge" (5.2.244–45). These men, like Shylock, "*crave* the law" (4.1.206, emphasis added).[24]

Shylock's "egalitarianism" finds its closest counterpart in Emilia, Iago's wife in *Othello*. She explains to Desdemona the potential of wives to stray, to be unfaithful to their husbands:

---

[20] Lucius Annaeus Seneca, *Epistulae Morales*, trans. R. M. Gummere, in *Seneca in Ten Volumes*, Loeb Classical Library (Cambridge, MA: Harvard University Press, 1971), vol. 6, epistle 106, 219.

[21] Galen, *The Diagnosis and Cure of the Soul's Passions*, in *Galen on the Passions and Errors of the Soul*, trans. Paul W. Harkins (Columbus: Ohio State University Press, 1963), 53.

[22] Anthony Munday, trans., *The knowledge of a mans owne selfe by P. De Mornay* (London, 1602), 126–27.

[23] Quoted in Harris, *Foreign Bodies*, 101.

[24] To "crave" in early modern English means to ask or demand but may, more specifically, denote the demands of physical appetite; *Oxford English Dictionary*, s.v. "crave."

Why, we have galls; and though we have some grace,
Yet we have some revenge. Let husbands know
Their wives have sense like them; they see, and smell,
And have their palates both for sweet and sour,
As husbands have.

.   .   .   .   .   .

And have we not affections,
Desires for sport, and frailty, as men have?
Then let them use us well; else let them know,
The ills we do, their ills instruct us so.
(4.3.92–6.100–103)

The similarity of Emilia's speech to Shylock's is striking: she too exploits the case for an analogy between human bodies—here, men and women's bodies instead of Jewish and Christian bodies—to explain the source of her "ills" and, moreover, a woman's "natural" inclination toward revenge. Clearly, Shylock is not alone in associating vengeance with nature and in making common nature, an equality of bodies, the grounds for a reciprocation of desires. There is evidence, indeed, that Shakespeare conceived gender relations as he did race relations—in terms of the measure of men and women's bodies against one another.[25]

The mid-sixteenth-century Protestant apologist Richard Hooker, expounding "that Law of Nature whereby human actions are framed," perhaps most clearly articulates the way that "natural equality" entails a fellow feeling by which we need, again, by *nature*, to experience the same pleasures as others as well as the same pains. His passage is important enough to cite at length:

---

[25] Like his notion of race relations, Shakespeare's "gender relations"—the measure of men and women against one another—may also be contingent on a proportionality of male and female body parts within and among people. Shakespeare's sonnet 20 sets up an intriguing paradigm: as the sonnet constructs the "master mistress" of the poet's passion, he or she has three female parts ("a woman's face," "a woman's gentle heart," and a woman's eyes) and one male "addition." Part for part, the sonnet portrays a person who is three parts female to one part male, and it is these specific proportions that create the conditions for the nature of the relationship between the poet and his beloved. Sonnet 20's model of "proportional" gender suggests a very different configuration of the relations between men and women than the "one body" model theorized by Thomas Laquer, according to which Renaissance man is the measure of all women. See Thomas Laquer, *Making Sex: Body and Gender from the Greeks to Freud* (Cambridge, MA: Harvard University Press, 1990).

[T]he like natural inducement hath brought men to know that it is their duty no less to love others than themselves. For seeing those things which are equal must needs all have one measure; if I cannot but wish to receive all good, even as much at every man's hand as any man can wish unto his own soul, how should I look to have any part of my desire herein satisfied, unless myself be careful to satisfy the like desire which is undoubtedly in other men, we all being of one and the same nature? To have any thing offered them repugnant to this desire must needs in all respects grieve them as much as me: so that if I do harm I must look to suffer; there being no reason that others should shew greater measure of love to me than they have by me shewed unto them. My desire therefore to be loved of my equals in nature as much as possible may be, imposeth upon me a natural duty of bearing to themward fully the like affection. From which relation of equality between ourselves and them that are as ourselves, what several rules and canons natural Reason hath drawn for the direction of life no man is ignorant; as namely, "That because we would take no harm, we must therefore do none."[26]

Hooker's theory of equality draws on ideas of natural law and "natural Reason" but also directly on the New Testament, specifically the words of Jesus: "Whatsoeuer ye wolde that men shulde do to you, euen so do ye to them" (Matt. 7:12; cf. Luke 6:31). Jesus affirms the significance of his injunction by calling it "the law" itself: "[T]his is the Law and the Prophetes" (Matt. 7:12). The Book of Common Prayer also included a part of Matthew 7:12 as the answer to the Catechism question "What is thy duty towards thy neighbor?"[27] Yet Hooker ultimately identifies the source of the Golden Rule not in theological doctrine but in the "equality between ourselves and them that are as ourselves." Just as important, Hooker's idea that "those things which are equal must needs all have one measure" leads him to acknowledge the natural basis of revenge: "[S]o that if I do harm I must look to suffer." Because we assess ourselves as "equals in nature," according to Hooker, we naturally exchange good for

---

[26] Richard Hooker, *Of the Laws of Ecclesiastical Polity*, in *The Works of Mr. Richard Hooker*, 3 vols., ed. John Keble (Oxford: Clarendon Press, 1888), 1:231.
[27] *The Book of Common Prayer, 1559: The Elizabethan Prayer Book*, ed. John E. Booty (Washington, DC: Folger Shakespeare Library, 1976), 286.

good, grief for grief, measure for measure. Both the Renaissance sciences of the body and the persistence of the principle of "natural equality" in literary, theological, and political discourse meant that the Old Testament law of retaliatory justice, "Thou shalt give life for life, eye for eye, tooth for tooth, hand for hand, foot for foot, burning for burning, wound for wound, stripe for stripe," was not so easily displaced by a New Testament code of "turning the other cheek." In the Renaissance, after all, both revenge and mercy—though posited as contrary ethical codes, one Old Testament, one New—were deemed instinctive responses to the relationship of one's self, one's own suffering, to others.

Shylock's need to project his own suffering onto Antonio, then, is not just about Old Testament law; it's about how Shylock, as a man, naturally feels. When he learns of his daughter's flight with Lorenzo, he grieves that he has no revenge "but what lights a' my shoulders, no sighs but a' my breathing, no tears but a' my shedding" (3.1.94–96). He grieves, in other words, that all the pain is his own; "satisfaction" of these feelings would affect Antonio's shoulders, Antonio's sighs, Antonio's tears, in kind. Shylock's speeches juxtapose expressions of his own pain with exclamations of how he will cause like pain in Antonio: "Thou stickst a dagger in me. . . . I'll plague him, I'll torture him. . . . Thou torturest me, Tubal. . . . I will have the heart of him if he forfeit" (3.1.110, 116, 120, 127). Iago too will describe his grief, jealousy, in terms of a "gnawing" of his "inwards," which can only be cured by getting "even" with Othello's body, by having Othello "eaten up" with the same passion. There is nothing metaphorical, for Shakespeare and many other Renaissance dramatists, about speaking of revenge as a bodily need to inflict that disease onto another's body: "I will live / Only to numb some others' cursed blood / With the dead palsy of like misery."[28] However immoral, Renaissance revenge is often predicated on a natural identification with other bodies as both cause and cure of pain.

The trouble with Shylock's famous speech on natural equality is not that it's illogical, sophistical, or "inhuman." The trouble is that Shylock, by the standards of Shakespeare's Venice, is *not* equal by nature to Antonio, however "much kindness in the Jew" there is, in terms of dimensions, organs, senses, and affections. The outcome of Shylock's trial

---

[28] John Marston, *Antonio's Revenge*, ed. Reavely Gair (Manchester: Manchester University Press, 1999), 2.2.71–72.

turns, finally, on a "pound of flesh," that is, on the relatively slight but ultimately critical *measure* of another's body in relation to his own.

> "[L]et the forfeit / Be nominated for an equal pound /
> Of your fair flesh"
> —SHYLOCK (1.3.148–49)

Shylock's trial makes literal the ancient and enduring metaphor of the "scales" of justice, revealing justice as an act of human measurement, the "weighing" of men, as Portia asks, "Are there balance here to weigh / The flesh?"[29] Generations of readers have admired Portia for her cleverness in turning the tables on Shylock's insistence on the letter of his bond, which granted him "an equal pound" of "[Antonio's] fair flesh" should the latter forfeit on the moneylender's loan. Noting that Shylock's "words expressly are 'a pound of flesh' " (4.1.307), Portia challenges the Jew to cut not "less nor more / But a just pound":

> If thou tak'st more
> Or less than a just pound, be it but so much
> As makes it light or heavy in the substance
> Or the division of the twentieth part
> Of one poor scruple, nay, if the scale do turn
> But in the estimation of a hair,
> Thou diest, and all thy goods are confiscate.
> (4.1.326–32)

Although there is no doubt that Portia is trading on Shylock's Old Testament literalism, she is also exposing an error in his calculations. It's not just the "letter" of his bond but its "number"—the impossibility of Shylock gauging a precise amount of Antonio's flesh—that finally makes the merchant's forfeit irredeemable. His revenge on Antonio is averted, the play's potential tragedy deflected, by the "estimation of a hair"; whatever the scales can accomplish, Shylock cannot possibly take an accurate mea-

---

[29] See chapter 5 for a fuller discussion of the rhetoric of measurement in Renaissance legal discourse.

sure of Antonio's body, properly value a "hair" nor any other part be-
longing, in this case, to a Christian man.[30]

In setting the forfeit of his bond with Antonio at "a weight of carrion
flesh," Shylock had disingenuously reminded Bassanio and Antonio of
how little he was asking, in terms of what the market would bear: "A
pound of man's flesh taken from a man / Is not so estimable, profitable
neither, / As flesh of muttons, beefs, or goats" (1.3.165–67). "Market
value," however, is clearly not Shylock's criterion of assessment. He specifies an "equal pound" of Antonio's flesh; "equal" may simply mean
"even" or "just" in a general (unquantified) sense here. But as James
Shapiro has detailed with evidence from Shakespeare's literary and cultural sources, Shylock's desire for a pound of Antonio's flesh suggests a
desire for a *particular* pound of the man. Apparently Shakespeare's
sources weighed a man's penis in at precisely that amount: Shakespeare's
play follows Alexander Sylvayn's *The Orator* in representing a Jew who
wonders if he "should cut off [a man's] privy members, supposing that
the same would altogether weigh a just pound?"[31] In early modern accounts of circumcision, including accounts of adults who "turn" Jewish
or Muslim, this ritual cutting was frequently "conflated and confused
with the idea of castration."[32] Certainly, Shakespeare repeatedly reminds
us of the possibility of adding or removing a penis from the human body
throughout the play, especially in the jests surrounding Portia and Ner-
issa's "turning men." They joke of feigning an "addition" to their bodies:
"Why, shall we turn to men?" "[T]hey shall think we are accomplished /
With that we lack" (3.4.78, 61–62). When they expose themselves at the

---

[30] The "estimation" of human value is attempted, often unsuccessfully, throughout
Merchant. The suitors who come in quest of Portia are challenged to evaluate not only
the respective value of the three caskets but their own worth in relation to them:
"Who chooses me shall get as much as he deserves" (2.7.7). Morocco, among them,
pauses to "weigh [his] value with an even hand" (2.7.23–25). Losers of the casket
game learn that weighing oneself "evenly" is no simple task, as Aragon concludes,
"Let none presume / To wear an undeserved dignity" (2.9.39–40). Antonio urges Bassanio to give up Portia's ring to the worthy judge who's saved his life: "Let his deserv-
ings and my love withal / Be valu'd 'gainst your wive's commandment (4.2.450–51).
Bassanio, in turn, accounts Antonio's survival more precious than all other life and living: "[L]ife itself, my wife, and all the world, / Are not with me esteem'd above thy
life" (4.1.284–85). The romantic plot of *Merchant* is designed to ensure that Portia be
"nothing undervalu'd" (1.1.165) compared with others.

[31] Quoted in James Shapiro, *Shakespeare and the Jews* (New York: Columbia University
Press, 1996), 121.

[32] Vitkus, *Three Turk Plays*, 5.

end of the play, Gratiano threatens, in jest of course, to "mar the young clerk's pen" (5.1.237). If Shapiro is right, the pound of flesh that Shylock wants of Antonio is just what many of Shakespeare's characters believe would make up, in material, quantified terms, the difference between man and woman, Christian and Jew. Shylock's "revenge" thus represents a desire to circumcise his Christian enemy, to make Antonio's Christian body commensurate with his own, to have him "turn Jew."

From a Christian perspective, it's obviously "unkind" of Shylock to want to do this; it proves, as the Duke fears, that he has not a "*dram* of mercy" in him (4.1.6, emphasis added). But what of the reverse? Is it kindness, in Shakespeare's terms, to turn a Jew Christian? Whatever the motives of Renaissance Christian evangelism, is such a turning possible? The Inquisitions of Spain, Portugal, and Italy doubted it, and some early modern Englishmen concurred that "Jewishness" simply cannot "be washed from [the Jewish body] with the sacred tincture of baptism."[33] Yet Christian conversion, from a theological perspective, had always promised a metamorphosis of "old" natures, first through self-denunciation (a rejection of one's old faith), but finally, and above all, through the ritual cleansing of baptism. The Elizabethan Book of Common Prayer prescribed a service based on Paul's assertion that those baptized in Christ's name are "regenerate" and born anew. Through baptism, the liturgy goes, "the Old Adam" within dies to give way to a "new man" who is then "*grafted onto the body* of Christ's congregation" (emphasis added).[34] Baptism was also the chief means by which the conversion of all peoples, culminating at the end of human history with the conversion of the Jews, was to be accomplished in this world. Sixteenth-century accounts of such conversions, such as *The Baptizing of a Turke* (1586), a sermon written on the occasion of the conversion of "Chinano, a Black Moor and Spanish captive," invariably rehearse the rhetoric of changed bodies and celebrate the convert's new "lineage": "This silly Turke and poore Saracen [was] by Baptisme . . . sealed up for the childe of God, and inheritour of the kingdome of heauen."[35] Yet for all that sixteenth-century baptism promised in the way of fully transforming the bodies of men, Renaissance writers were not always sure whether Muslims, or Jews for that matter, were equal to that degree of change. For many, the question re-

[33] Quoted in Shapiro, *Shakespeare and the Jews*, 19.
[34] *Book of Common Prayer*, 274–75.
[35] Meredith Hanmer, *The Baptizing of a Turke* (London, 1586), "Epistle Dedicatorie."

mained: Can conversion equate the natures of men born so incommen-
surate?

Shakespeare's plays are filled with characters who assert the transfor-
mative power of conversion, including perhaps Shylock, who sees his
daughter Jessica's conversion, her marriage to a Christian, as a rebellion
against nature, against his own body: "My own flesh and blood to rebel!"
Notably, Salerio offers a salacious interpretation of Shylock's exclama-
tion: "Out upon it, old carrion, rebels it at these years?" (3.1.34–35).
That Salerio deliberately misunderstands Shylock's remark about his
"own flesh" as a reference to his penis is not incidental, given one bodily
site of Shylock's "Jewishness." Yet Salerio goes on to assert the natural difference between father and daughter: "There is more difference between thy flesh and hers than between jet and ivory, more between your
bloods than there is between red wine and Rhenish" (3.1.38–42).
Lorenzo agrees that Jessica is different *by nature* than her father, despite
her being "issue to a faithless Jew" (2.4.37), and makes a point of referring to her as "gentle Jessica" (19), rhetorically baptizing her with a
Christian name ("gentile"). Launcelot, on the other hand, is more
doubtful, arguing that only a "bastard hope" (3.5.7), the chance that
she's not really her father's daughter, could save her soul, exchange her
Jewish nature for a gentile one. When Launcelot tells Lorenzo that he is
"no good member of the commonwealth" (34–35) for converting a Jew,
Lorenzo replies that he's a better Venetian than Launcelot, who has im-
pregnated a Moor. From the perspective of some of Shakespeare's characters, the marriage of Christian and Jew is form of miscegenation, a rebellion of one's own flesh against the rules of nature. There's an implicit
question here too as to whether the union of Christian and Muslim represents an even greater "misrule."[36]

We would, however, merely be mimicking Shakespeare's characters
simply to conclude that conversion either "works" or it doesn't in his
plays, that his characters either shed their old natures and assume new
ones or stay exactly as they were before. As Jessica's example, and later,
Othello's, suggests, Jews and Muslims may come to approximate Christian "kindness," even as they retain "parts" of their former natures; in
other words, changes in nature may be partial ones. Jessica is not and

---

[36] See Adelman, "Her Father's Blood." Adelman sees the problem of Jessica's "conversion" as a key to understanding the relationship of "race" and "nation" in the Renaissance.

probably never was "equal by nature" to her father, yet she shows signs of threatening to lapse again into a kind of "infidelity." In their last scene together, as has often been noted, Lorenzo and Jessica mockingly refer over and over again to their "vows of faith, / And never a true one" (5.1.19–20). Lorenzo, responding to Jessica's appreciation of the musical harmonies around them, compares her reaction to one he might expect of wild beasts:

> The reason is, your spirits are attentive;
> For do but note a wild and wanton herd
> Or race of youthful and unhandled colts,
> Fetching mad bounds, bellowing and neighing loud,
> Which is the hot condition of their blood,
>
> . . . .
>
> You shall perceive them make a mutual stand
> Their savage eyes turn'd to a modest gaze,
> By the sweet power of music.
> (70–73, 77–79)

This passage has often been taken as an idealization of the social harmony of Belmont via a metaphor of music, but such an interpretation ignores the identification Lorenzo makes here between his new wife and that "race of youthful and unhandled colts" who can also appreciate a beautiful song. Implicitly, Jessica too descends of a "race" whose blood, according to Renaissance genealogies, is conditioned to wantonness, and if the more savage aspects of her nature are tamed here by music, the threat of their reemergence remains, just as it does for a wild beast: "[N]ought so stockish, hard, and full of rage, / But music *for the time* doth change his nature" (5.1.81–82, emphasis added). Jessica is *more* Christian than Shylock, but perhaps not wholly Christian for all that.

If Shakespeare calls the integrity and irrevocability of Jessica's conversion, which is unforced, into question here, how much more suspicious must he be of Shylock's compulsory one, which even the Inquisitions of Europe would have held invalid? It's hard to imagine that Shylock would entirely change his nature at the font (4.1.400); nothing in the play leads us to believe he can ever be exactly what Antonio is, "[a] *kinder* gentleman treads not the earth" (2.8.35, emphasis added). Given its likely failure, forced conversion, whether in the form of Shylock's attempt to claim a pound of flesh from Antonio or Antonio's attempt to baptize Shylock, is just another way of saying "revenge," or getting "*even.*"

Beyond the prospects for conversion, however, the idea of a human "equality by nature," the commensuration of bodies across the tribes of man, is the ultimate target of Shakespeare's inquisitions of human nature in *Merchant.* Part for body part, feeling for feeling, Shylock and Antonio can never be equal in the strictest, quantitative sense. If their "equality" cannot be absolute, however, then neither can their "inequality," despite what Lancelot and others believe about the difference in their flesh. "Nature" is not an absolute value in the *Merchant of Venice*; if it were, it would be impossible to explain why Shylock and Antonio, Jew and Christian, in certain ways have more in common in their passions and their griefs than, say, Bassanio and Antonio do (as evidenced, especially, by the moneylender's and the merchant's hatred for one another).[37] In relating Shylock and Antonio by a kinship of feeling, Shakespeare creates a new genealogy that competes with older or more familiar lines of "natural" affiliation.

For many Renaissance writers, the difference between the bodies of Christians and Jews might have seemed obvious enough. Lesser degrees of difference, however, even among men of the same "kind," were also acknowledged by Renaissance physicians and other close observers of human nature. Despite the common refrain in contemporary medical treatises, that "the life of all things is all one with it self in all persons," many take note of the "monstrosities" or anomalies that distinguish persons, physiologically, from one another.[38] Galen himself appealed to the unique nature of individual bodies in his effort to cure disease:

> It is evident that both the nature of the patient is to be considered, and also that there is a proper curation for euerie man, and yet furthermore . . . there is an ineffable propertie of euerie nature, neither comprehensible by the most exact knowledge. He is the best Phisition of euerie particular patient, which hath gotten the method, whereby he may discerne natures, and also coniecture

---

[37] Structurally, Shylock and Antonio also occupy analogous positions in the play as "forfeits" guaranteeing the happiness of others.

[38] Nicholas Culpeper, trans., "To the Reader," in *The Anatomy of the Body of Man* (London, 1653). Nancy G. Siraisi, "Vesalius and Human Diversity in *De humani corporis fabrica*," *Journal of the Warburg and Courtauld Institutes* 57 (1994): 60–88, quote at 68. Siraisi notes that Vesalius inherited Galen's belief in a "canonical" human body but believed that the purpose of dissection was additionally to discover "the difference of bodies" (71).

which are the proper remedies of euerie one. For it is an extreame madnesse, to judge that there is a common curation of all men. As though they should cure a universall and not a particular man.

Galen adds, "[I]f I knew how to find out exactly euerie private nature, I would think my self to be such a one, as I conceiue in my minde, was Aescalapius." The distinctive feature of "euerie priuate nature" is called an "idiosyncracy," as Galen explains, "[T]he common sort call it in greeke Idiosyncrasian, and all they confesse it to be incomprehensible."[39] The word "idiosyncrasy" entered early modern English, as a medical term, in 1604.[40] For all their belief in a universal human nature, Renaissance writers increasingly ask of each other's grief, as Gertrude does of Hamlet's, "Why seems it so particular with thee?" (1.2.75).

For Shakespeare, the problem of "idiosyncrasy"—the diagnosis of private "natures"—is an epistemological one: how can we know the pain of another? More pragmatically, it casts some doubt on the objectives of revenge: how is it possible to get "even" with another when one's griefs are particular to oneself? Matthew Poole, in his commentary on the "law of retaliation" as described in Exodus, pointed out the practical difficulties in achieving the revenge of "an eye for an eye," the same problem facing Shylock: "[H]ow could a Wound be made neither bigger nor less than that which he [the original assailant] inflicted?"[41] It is no wonder that revenge, the proportioning of grief for grief, never seems satisfactorily to "cure" the pain of Shakespeare's avengers; something always remains as a natural lack or excess.[42] For better or for worse, Hamlet's difference, like Shylock's, survives his surrender to what he is compelled to become by the exigencies of revenge. These men remain more (or less) than those they seek to destroy. Shakespeare did not judge revenge to be right or wrong, perhaps, so much as unsatisfying, unequal to the demands of human desire.

---

[39] Galen, *Certain workes of Galens, called Methodus medendi*, trans. T. Gale (London: 1585), sigs. G7v, G8v. The first-century Roman physician Celsus also stressed the need to understand patients' individual features or "idiosyncracies."

[40] *Oxford English Dictionary*, s.v. "idiosyncrasy."

[41] Matthew Poole, *Annotations upon the Holy Bible* (London, 1683), vol. 1, sig. T5v.

[42] Hamlet is not alone among Shakespeare's characters in experiencing his loss as particular to him; many consider their grief as sui generis. Compare Brabantio, "my particular grief / Is of so flood-gate and o'erbearing nature / That it engluts and swallows other sorrows, / And it is still itself" (*Othello*, 1.3.55–58).

Shakespeare's Brutus, in *Julius Caesar*, referring to his professed friendship for the man he's about to murder, mourns his recognition that *omne simile non est idem*, "every like is not the same" (2.2.128). Shakespeare understood the *desire* for reciprocity, for an equality of feeling, to be something we naturally share, including the desire to get "even" with others. He also understood, however, that satisfying that desire may be impossible. Shylock got it right when he enumerated the similarities between Jewish and Christian bodies, but "every like is not the same," as Shylock himself secretly knows well enough. The Jew offers "kind" to Antonio, but he doesn't mean it; he is fully aware of the degree of difference that makes his offer, and his apparent Christian "kindness," a fraud. Although the play never explicitly mentions the Inquisition, or the imagined threat of the converso in the early modern imagination, *The Merchant of Venice* confesses a certain sympathy for an early modern distrust of the alien bodies of Jews.

Shakespeare's judgment of Shylock is not nearly that simple, however. If he sets up Antonio's nature as a measure for Shylock's, a standard by which to judge him, the reverse is also true. Shylock has feelings too, "natural" human affections that legitimize his assessment of his enemy. By Shakespeare's scales, the two must be weighed, can only be weighed, against one another. Neither Christian nor Jew represents, for Shakespeare, an exclusive, absolute human standard. Both, from Shakespeare's perspective, are "normal," including in the means through which they choose to pass judgments on others. They measure from themselves.

> Commend me to my kind lord.
> —DESDEMONA, on Othello (5.2.125)

> I never knew a Florentine more kind.
> —CASSIO, on Iago (3.1.40)

Shakespeare's second Venice play returns to the kindness of two more strangers to Shakespeare's Venice, Othello and Iago, and considers what is between the Moor and his ensign, racially speaking. Reevaluating their relationship in the context of Shakespeare's first Venice play, featuring the Jew Shylock, and in the context of race relations among Jews and Muslims imagined in Renaissance literature more generally, I will argue something that I hope won't finally seem too outlandish—that Iago's nature, in relation to Othello's, partakes of just a little Jewishness, that Shakespeare's second Venetian villain is conceived to some degree as kin to Shylock, if less than equal by nature to him.

A complex nexus of racial terms has long been associated with Shakespeare's Othello. Along with explicitly classifying Othello as a Moor, the play also insinuates that he is some kind of Turk and, perhaps, some kind of Spaniard as well. Othello's psychological transformation corresponds to the play's historical events, the wars of Venice against the Ottoman Empire, so that in the process Shakespeare sets up an analogy in which the danger of "turning Turk" applies to Cyprus as well as to the Moorish general sent to defend it. Othello makes that identification explicit in his final moments. Iago too, whatever kind of Italian he may be,[43] also relates himself to the Turk. When he jests with Desdemona about women's infidelity, he insists that all women are unchaste, swearing, "Nay, it is true, or else I am a Turk: / You rise to play and go to bed to work" (2.1.114–15). Because what he says about women is not true—Desdemona's example contradicts his assertion, since she is faithful to Othello—Iago exposes something "Turkish" in his own nature, just as Othello does.

Many scholars have also discerned something distinctively Spanish about Iago. Shakespeare's primary literary source for the character, Giraldi Cinthio's *Hecatommitthi*, calls him only "the Ensign" but indicates a foreign birth at the end of the story when we're told that "the Ensign returned to his own country" [ando l'Alfieri alla sua patria].[44] Shakespeare seems to have followed Cinthio in suggesting that Othello's lieutenant is something other than a native Venetian. In assigning him the name Iago, a Spanish name rather than an Italian one, Shakespeare may well have been thinking of Santiago, Saint James, the patron saint of Spain, and his contemporary association with the Reconquista or military expulsion of the Moors from that nation. The Spanish believed that Santiago led them in their defeat of the Moors, which earned him the epithet Santiago Matamoros—the "moorslayer." Iago's "Spanish" nature, per-

---

[43] Mary Floyd Wilson asserts that "Iago is the epitome of the Italianate Italian" (*Ethnicity and Race*, 142); she bases her case on the evidence of Iago's "Italianate" jealousy. As I show in this chapter, however, "jealousy" is also often identified as a passion pertaining to Jews and Muslims. I have no doubt that Shakespeare's Iago is an Italian—but, crucially, he is an Italian only in part.

[44] Giraldi Cinthio, *The Moor of Venice*, trans. John Edward Taylor (New York: AMS Press, 1972), 35.

haps, shows Shakespeare's subscription to Renaissance English anti-Spanish sentiment or the Black Legend, including propaganda directed against the Inquisition.[45] Iago not only has a Spanish name but seems to speak Spanish, crying "Diablo!" as he pretends to break up the brawl he himself incites between Cassio and Roderigo (2.3.161).[46] If Iago is a Spaniard, however, Othello, not incidentally, probably is as well: Leo Africanus, author of *History and Description of Africa* and source for the character of Othello, was a converted Moor, born in Granada.[47] Contemporary literature provided many other examples of Spanish Moors.[48] A key source for Shakespeare's *Othello, Lust's Dominion* (perhaps an alternate title for the lost play *The Spanish Moor's Tragedy*) centers on the story of Eleazar, son of the former king of Barbary, a Moorish warrior who boasts of many "victories / Achieved against the Turkish Ottoman" on behalf of Spain.[49] If Shakespeare imagined Othello as a Spanish Moor as well, this would explain how his protagonist comes by the "sword of Spain" he draws to kill himself (5.2.253).

Moor, Spaniard, Turk. These terms, as has often been noted, frequently occur together in Renaissance conceptualizations of race. Yet one term is missing from this set of Renaissance racial relations: the Jew is so often found, throughout Renaissance texts, in the same semantic field. I would like now to present evidence that Shakespeare's Iago de-

---

[45] For the relationship between *Othello* and the Black Legend, see Andrew Hadfield, "Race in Othello: *The Description of Africa* and the Black Legend," *Notes & Queries* 45 (September 1998): 336–38; Eric Griffin, "Un-Sainting James: or, Othello and the 'Spanish Spirits' of Shakespeare's Globe," *Representations* 62 (Spring 1998): 58–99; Peter Moore, "Shakespeare's Iago and Santiago Matamoros," *Notes & Queries* 43 (June 1996): 162–63; and John A. Rea, "Iago," *Names: A Journal of Onomastics* 34.1 (March 1986): 97–98. For a general study, see William S. Maltby, *The Black Legend in England: The Development of Anti-Spanish Sentiment, 1558–1660* (Durham, NC: Duke University Press, 1971).

[46] For a recent discussion of Iago as a Spaniard, see Griffin, "Un-Sainting James." Among the evidence Griffin cites is a contemporary ballad about Shakespeare's play that speaks of Iago as a "false Spaniard" (67). Griffin sees Iago's antipathy toward Othello as that of an "Old Christian" toward a "New" one (80).

[47] Leo Africanus, *The History and Description of Africa*, 3 vols., trans. John Pory (1600), ed. Robert Brown (New York: Burt Franklin, 1963).

[48] For the case that Othello too is a Spaniard, see Barbara Everett, "'Spanish' Othello: The Making of Shakespeare's Moor," *Shakespeare Survey* 35 (1982): 101–12.

[49] *Lust's Dominion*, 2.3, p. 76. See Bekkaoui's edition for evidence that this may be an alternate title for *The Spanish Moor's Tragedy*, a lost work recorded by Philip Henslowe and written by Thomas Dekker, William Haughton, and John Day. Bekkaoui suggests 1599–1600 as a date for *Lust's Dominion*.

rives not only from the Spaniard but from the Jew, especially, the Renaissance Jew who turns Turk, in conformance with his nature, on behalf of the "general enemy Ottoman" (1.3.49). Shakespeare wrote a play about a Jew of Venice and then, a few years later, one about a Moor of Venice; where there's a Jew, for many Renaissance writers, a Muslim isn't far behind. And vice versa: a distant relative has followed Othello to Cyprus.[50]

Jews and Muslims, no doubt, are closely affined in the Renaissance imagination. The parallel in their fortunes in sixteenth-century Spain explains this in part. In Spain, the Jewish question and the Muslim question were posed as related concerns, and the two groups shared similar fates over the course of the sixteenth century. The year 1492 not only marked what was supposed to be the final solution to the Jewish problem—the expulsion of the Jews from the Peninsula—but also the end of the Reconquista of Moorish Spain with the seizure of Granada. The final solution to the Moorish question came a century later, in 1609, when they too were banished from the realm. The official rationale for the eviction of the Jews, set forward in the Spanish monarchs' edict of expulsion, was to protect New Christians, to end "the great harm suffered by Christians [i.e., conversos] from the contact, intercourse and communication which they have with the Jews, who always attempt in various ways to seduce faithful Christians."[51] Yet in one of the great historical ironies of the period, the Inquisition itself targeted the same New Christians the expulsion sought to protect. Conversos were suspected of secret judaizing, of retaining not only their original observances but their original "natures" as well. What protests there were from Spanish Christians often echoed the words of Saint Paul: "[All Christians are] under one baptism, formed under the law into one body, so that whether Jew, Greek or Gentile we are regenerated by baptism and made into new men. From

---

[50] Rather than exemplifying a marked racial or ethnic type, Iago has sometimes been described as the "Turk" within all of us, a repository of inner demons that are dangerously projected outward onto others. According to this reading, Iago himself is not racialized—his own identity is null or emptied out via his projections; rather, he is the source of racialized thinking, once who marks others even as he resists any such definition himself. See Janet Adelman, "Iago's Alter-Ego: Race as Projection in Othello," *Shakespeare Quarterly* 48.2 (1997): 125–44. I do not think that Iago is "unknowable" (144) or lacking a (racial) identity, or rather I believe that Iago is no more and no less "knowable" than Othello is in this regard.

[51] Quoted in Henry Kamen, *The Spanish Inquisition: A Historical Revision* (New Haven: Yale University Press, 1997), 20.

which it is obvious how culpable are those who, forgetting the purity of
the law of the gospel, create different lineages, some calling themselves
Old Christians and others calling themselves New Christians or conversos."[52] Yet the Marranos, as New Christians of Jewish descent were
known, were persecuted despite their baptism and considered the same
"old" men and women as their Jewish lineages determined them to be.

By 1501, in turn, the Moors or Mudejares (as they were known in
Spanish) of Granada had undergone baptism by decree, as had those in
many other regions of Spain. These New Christians too were given a new
name, Moriscos. By 1526 the Muslim religion no longer officially existed
in Spain, but a new century of rebellions led to a call to resolve the
Moorish question, once and for all, despite some reservations: "It would
be terrible to drive baptized people into Barbary and thus force them to
turn Muslim."[53] Although this didn't take place until 1609, five years
after Shakespeare wrote *Othello*, the prospect of expelling the Moors was
already in view. *Lust's Dominion* ends with the Spanish king pronouncing
against Eleazar, the Moor, and all his kind: "And for the Barbarous Moor
and his black train, / Let all the Moors be banished from Spain." Queen
Elizabeth herself had issued a decree to deport any Moors seeking
refuge from the Inquisition:

> [T]he Queen's Majesty, tendering the good and welfare of her own
> natural subjects greatly distressed in these times of dearth, is
> highly discontented to understand the great numbers of Negars and
> Blackammoors which (as she is informed) are crept into this
> realm since the troubles between Her Highness and the King of
> Spain, who are fostered and relieved here to the great annoyance
> of her own liege people and want relief which those people con-
> sume; as also for that the most of them are infidels, having no un-
> derstanding of Christ or his Gospel.[54]

The Spanish—perhaps the English too—dealt with the Moors in ways
that often simply repeated earlier steps taken with the Jews.[55]

---

[52] Ibid., 35.
[53] Ibid., 227.
[54] *Lust's Dominion*, 5.6., p. 150; p. 18.
[55] Kamen, *Spanish Inquisition*, 216.

European accounts of the Spanish Inquisition also helped establish the idea that Jews and Muslims were kin by nature rather than just by circumstance. It is often said that sixteenth-century accounts of the Inquisition ignored the plight of Jews and Moors, focusing almost exclusively on the threat that Spain posed to European Protestants. Early modern writers generally trace the sources of the Inquisition to efforts to suppress Judaism and Islam, or to root out apostasy among New Christians. John Foxe, for example, notes that the Inquisition was originally "instituted against the Jews" who "after their baptism maintained again their own ceremonies."[56] Reginald Gonzalvus Montanus's *Sanctae Inquisitionis Hispanicae Artes* (1567), one of the chief sources of sixteenth-century Protestant propaganda regarding the Inquisition, begins with a short history of the Reconquista and the "diabolical machinations" of the Jews, Moors, and conversos against the Spanish crown. *The Tragicall Historie of the warres in the Low Countries* notes that the Inquisition gained ascendancy after the Jews crucified a Christian child on Palm Sunday, 1475.[57] According to Montanus, Foxe, and others, however, the original, noble goals of the Inquisition were perverted when fanatics gained control of it and used it to enforce uniformity among Christians. Most contemporary scholars concur that the Protestant interest in Jews and Moors ends here.

That's not the case at all, however. In many of these accounts, Jews and Muslims turn out to be the very perpetrators of the persecution originally aimed at them, and moreover, they always were. How can they be represented as both tortured and torturer? The answer lies in blood. Mayerne Turquet, in his *Historie of Spaine* (1583), explains that the Spanish Jews' "profession, either true or fained, of Christian religion" led to the mixing of Jewish and Spanish blood to such an extent that in the "process of time the noble families of Spain, allying themselves by marriage to that race, did wholly contaminate and pollute themselves both in blood and belief."[58] Similarly, Robert Ashley, translator of the French treatise *A Comparison of the English and Spanish Nation* (1589), recounts the history of Moorish rule in Spain and concludes: "[I]f of good right the Gothes and Vandales, are counted cruell, the Moores perfidious and

---

[56] Adelman, "Her Father's Blood," notes the tendency of these texts to affiliate Spaniards with Jews and Moors through their "bloodlines" (12).

[57] Shapiro, *Shakespeare and the Jews*, 17.

[58] Ibid., 18.

reuengefull, the Saracens proud, and villanous in their manner of liuing, I pray you what humanitie, what faith . . . may wee thinke to finde amongest this scumme of Barbarians?" To consider "the naturall disposition of the Spaniarde," Ashley translates, is not only to recognize the "barbarian" strain in him, but also how much, "resembling Iuie [Jews] . . . [he] hath alwaies made those to feele the most pernitious effects of [their] most hurtfull malice."[59] William of Orange, decrying the brutality of the Spanish in his *Apology* of 1581, states, "I will no more wonder at that which all the world believeth: to wit, that the greatest part of the Spaniards, and especially those that count themselves noblemen, are of the blood of the Moors and Jews."[60] Lewis Lewkenor's *Discourse of the Usage of the English Fugitives by the Spaniard* (1595) reviles the Spanish as "the most base, wicked, proud, and cruellest nation that liueth," noting the sources of that cruelty:

> [They are] a nation not yet fully an hundred yeeres since wholie they receiued Christianitie, and as yet are in their hearts a greate number of them, pagans & moores, from profession of which, they are onely helde by the seuere bridle of the sanguinarye and most cruell Heathen Inquisition. [T]hat Paganisme of theirs, which in theyr profession they dare not shew, they do in theyr tyranie, blasphemie, sodomie, cruelties, murther, adulterie, and other abhominations sufficiently discover.[61]

Finally, Antonio Perez, writing of "the naturall disposition of these [the Spanish] nations" in his *Treatise Paranaenetical* (1598), notes the Spanish tendency to "turn Turk" at the nearest opportunity: "The Castilians being in the countrey of the Saracens, or of the Turks, they do soone and verie easily denie the Christian faith, and do turne altogether Turk and Infidell." He too offers a "natural" explanation: "[T]hey are descended of the Vandals, of the Iewes, and of the Moores."[62] The "truth" about Spanish blood was validated by Pope Paul IV, who in 1556 referred to

---

[59] Robert Ashley, trans., *A Comparison of the English and Spanish Nation* (London, 1589), 19, 41.

[60] Quoted in Kamen, *Spanish Inquisition*, 310.

[61] Sir Lewis Lewkenor, *Discourse of the Usage of the English Fugitives by the Spaniard* (London, 1595), sig. E4.

[62] Antonio Perez, *A Treatise Paranaenetical* (London, 1598), 24.

them as "that breed of Moors and Jews, those dregs of the earth"; indeed, the Spanish royal family itself was suspected of having Jewish blood.[63] By a twist of genealogical irony (though, to be sure, an irony that seems to be lost on the authors of these works), the Spanish turn out to be Jews and Muslims themselves, by "sanguinarye" descent, the archenemies of Christendom assuming yet another new disguise. In discovering the Spanish as secret Jews and Moors, the Black Legend exposes its own sympathy with the Inquisition it purports to condemn.

Yet the idea of what historians have called a "Jewish-Muslim symbiosis" goes back much farther than the sixteenth century. It originates with the earliest Christian accounts of the birth of Islam. Mohammed was said to have had an evil Jew as his closest adviser and to have incorporated Jewish beliefs (as well as Christian ones) into the dangerous faith he promulgated, as a strategy for converting members of those religions. Both Jews and Muslims, from the Middle Ages through the early modern period, were often deemed descendants of the devil and always unyielding enemies of Christendom. Both are commonly reviled as "slaves" or "dogs" (for example, the Christians call Eleazar "a Moor, a Devil, / A slave of Barbary, a dog").[64] Not incidentally, it was commonly noted that both Jews and Muslims had bodies scaled back by circumcision.

These relations are nowhere more conscientiously and creatively imagined than in Renaissance drama, where Jew, Moor, and Turk so often appear together, cheek by jowl. The drama of the period repeatedly enacts their "relations," creating further, literary lines of descent. Martin Luther's literary allusions to the alliance between Jews and Turks may the earliest of the sixteenth century.[65] When the Italian Arnoldo, in John Fletcher and Philip Massinger's *The Custom of the Country* (written 1619), takes money of a well-seeming Jew, his countryman Rutillio comments,

> Sure, thy good angel is a Jew and comes
> In his own shape to help thee. I could wish now
> Mine would appear so, like a Turk.[66]

[63] Quoted in Everett, "'Spanish' Othello," 105.
[64] *Lust's Dominion*, 1.2, p. 55.
[65] Allan Harris Cutler and Helen Elmquist Cutler, *The Jew as Ally of the Muslim: Medieval Roots of Anti-Semitism* (Notre Dame: University of Notre Dame Press, 1986), 117.
[66] John Fletcher and Philip Massinger, *The Custom of the Country*, ed. Nick de Somogyi (Cambridge: Routledge, 1999), 2.3.60–62.

In Christopher Marlowe's *Jew of Malta,* a source for *Merchant of Venice,* Jew
and Turk meet again. This play, like Shakespeare's, is set in an interna-
tional center of trade over which Turks and Christians vie for power.
When the Christians need money for tribute to the Turk, they seize upon
the Jew Barabas's goods, threatening that he'll have to "straight become
a Christian" if he denies to pay.[67] Barabas swears he will be "no conver-
tite" (1.2.86) and rejects the Christian's insistence on his "inherent
sinne" (1.2.109) as a Jew:

> Some Jewes are wicked, as all Christians are:
> But say the Tribe that I descended of
> Were all in generall cast away for sinne,
> Shall I be tryed for their transgression?
> The man that dealeth righteously shall live.
> (1.2.112–16)

Like Shylock, Barabas blames his double dealings on Christian example:
"This is the life we Jewes are us'd to lead; / And reason too, for Christians doe the like" (5.2.115–16). Jews and Moors on the Renaissance
stage characteristically claim the likeness of their blood and bodies, their
desires and demands, to those of Christians.

What may be most telling about Marlowe's play, as an account of the
relations between Jews and Muslims, is that Barabas murders his own
daughter, Abigail, and adopts the treacherous Turk Ithamore as his new
heir. It's as if Barabas recognizes the "natural" tie between them, as he describes it: "[W]e are villaines both: / Both circumcized, we hate Christians both" (2.3.214–15). Abigail, among others, had recognized their
mutual "unkindness": "I perceive there is no love on earth, / Pitty in
Jewes, nor piety in Turkes" (3.3.48–49). It has long been recognized that
Barabas is a direct source for Shakespeare's Shylock, but he may be a
progenitor of Iago as well. Among other points the villains have in common, he too speaks Spanish, as he celebrates, for example, the "Her-
moso Placer de los Dineros" ("the beautiful pleasure of money"; 2.1.64).

---

[67] Christopher Marlowe, *The Jew of Malta,* in *The Complete Works of Christopher Marlowe,*
ed. Fredson Bowers (Cambridge: Cambridge University Press, 1973), vol. 1, 1.2.76.
All subsequent quotations from this play come from this edition and are cited parenthetically in the text.

Many Renaissance English plays set in the Ottoman Empire feature Jews who aid and abet the Turks—generally to their own advantage—including *Selimus, Emperor of the Turks* (1594), which stages the machinations, on Turkish territory, of an unregenerate Jew called Abraham. R.W.'s *Three Ladies of London* (1584), another source for Shakespeare's *Merchant*, features an Italian merchant named Mercadore who seeks to avoid paying back the three thousand ducats he owes the Jewish moneylender Gerontus by "turning Turk," since by Turkish law, "if any man forsake his faith, king, countrie, and become a Mahomet, / All debts are paide."[68] *The Travels of the Three English Brothers* (1607) by John Day, William Rowley, and George Wilkins, treats the sometimes historical adventures of the Sherley brothers among the Turks and Persians. The *Travels* is especially interesting for its explicit discussions of the meaning of racial and religious differences. When the Sophy of Persia asks Sir Anthony Sherley, "And what's the difference 'twixt us and you?" the Englishman responds, "None but the greatest" (i.e., the difference of their gods):

> All that makes up this earthly edifice
> By which we are called men is all alike.
> Each may be the other's anatomy;
> Our nerves, our arteries, our pipes of life,
> The motives of our senses all do move
> As of one axletree, our shapes alike.
> One workman made us all, and all offend
> That maker, all taste of interdicted sin.
>
> . . . .
>
> We live and die, suffer calamities,
> Are underlings to sickness, fire, famine, sword.
> We all are punished by the same hand and rod,
> Our sins are all alike; why not our God?[69]

[68] R.W., *The Three Ladies of London* (1584), ed. H. S. D. Mithal (New York: Garland, 1988), 2.1712–14.

[69] John Day, William Rowley, and George Wilkins, *The Travels of the Three English Brothers*, in *Three Renaissance Travel Plays*, ed. Anthony Parr (Manchester: Manchester University Press, 1995), 1.162–80. All subsequent quotations from this play come from this edition and are cited parenthetically in the text.

According to Sir Anthony, the natures of men are so alike that "[e]ach may be the other's anatomy"; their differences amount to "[o]nly art in a peculiar change / [Which] [e]ach country shapes as she best can piece them" (172–73). The Turkish beauty Dalibra finds no difference between their natural kinds either: "Strangers? I see no strangeness in them. The speak as well or, rather, better than our own countrymen" (3.22–23). Yet as Englishmen and Persians celebrate their commonalities, the evil Jew Zariph plots "[t]o taste a banquet all of Christian flesh" (9.23). Zariph, like Barabas, and like Shylock, judges his feelings by a Christian standard: "If we be curst we learnt of Christians / Who, like swine, crush one another's bones" (9.37–38), using the dictum of natural equality as a motive for vengeance.

Among dramatic works that try to correlate the natures of Jews and Muslims, the one that has the most in common with *Othello* is Robert Daborne's *A Christian Turned Turk*, published in 1612 but probably written between 1609 and 1611. It may be that Daborne was thinking of *Othello* or that the two plays are analogues, contemporaneously exploring the problem of the renegade or the Christian who "turns Turk." Either way, the similarities between these works suggest how Shakespeare might have imagined the nature of Iago's villainy. Daborne's drama focuses on two "notorious pirates," the historical John Ward and Simon Dansiker, and includes a scheming Jew named Benwash, who is in political and economic league with the Turks. Ward and Benwash have a lot in common, however: both "turn Turk" for a woman's sake.[70] The Jew converts to Islam to protect his Turkish wife from the presumed lasciviousness of Muslims, or as he puts it, to keep "[m]y bed free from these Mahometan dogs."[71] Benwash rationalizes his conversion:

If this religion were so damnable
As others make it, that God which owes the right,
Profaned by this, would soon destroy it quite.
(7.38–40)

At first Ward sees it differently, as he responds to the Jew:

---

[70] Vitkus, *Three Turk Plays*, 38.
[71] Robert Daborne, *A Christian Turn'd Turk*, in Vitkus, *Three Turk Plays*, sc. 6.76. All subsequent quotations from Daborne come from this edition and are cited parenthetically in the text.

That's easily answered: heaven is merciful.
By their destruction it should take all means
From giving possibility to their change.
(7.41–43)

Yet Ward himself soon turns renegade for love, deciding, expediently, that all religion is just "superstitious bondage" after all (7.203). There are many jokes made at his expense that center on his circumcision; a witness declares that he "saw him Turk it to the circumcision" (9.2), which by conventional Renaissance racial wisdom means he Jewed it there as well: "Marry, therein I heard he played the Jew with 'em" (3).

The most significant discussion about the differences among their kinds takes place between Agar, Benwash's Turkish wife, and Rabshake, Benwash's Jewish servant. When Rabshake describes the pirate Ward as a "reasonable handsome man of a Christian" (6.7–8), Agar asks, "Why? Doth religion move anything in the shapes of men?" (9). Rabshake replies,

> Altogether! What's the reason else that the Turk and Jew is trou-
> bled (for the most part) with gouty legs and fiery nose? To express
> their heart-burning. Whereas the puritan is a man of upright calf
> and clean nostril. (10–13)

For Rabshake, the difference between Turk and Jew on the one hand (note that these are considered of the same "nature") and Christian on the other hand is unmistakable, a matter of bodies ("the shapes of men"). When Rabshake is invited to turn Christian himself, however, he finds them similar enough:

> I turn Christian? They have Jew enough already amongst 'em. . . .
> I turn Christian? They shall have more charity amongst 'em first!
> They will devour one another as familiarly as pikes do gudgeons
> and with as much facility as Dutchmen do flapdragons. (24–27)

For Rabshake, Christians are already "Jews" in their taste for Christian flesh. Agar is horrified by Rabshake's imputation of cannibalism, "How? Eat up one another?" but the Jew insists to her it's true, "Aye, eat up one another!" (28–29).

Benwash's chief grievance from the start is jealousy (that is what motivates him to turn Turk in the first place). At first his jealousy, like Iago's, is invisible to those around him:

> You are of a noble mind, sir, courtly and high.
> It's want of merit that breeds jealousy,
> From which I know you clear.
> (6.65–67)

Benwash (like Iago) never seems sure of his own merit, however, and he ends up murdering his Turkish wife, who is seduced by another man. He is executed for the murder but feels himself blameless to the end, denouncing his conversion: "My wife proved false, untrue. / Bear witness, though I lived a Turk, I die a Jew" (16.212–13). The surviving Christians, including Ward (who ultimately comes to his senses and returns to his faith, although he kills himself for his error), declare that Benwash is no man: "It's / A name that bears too much of pity in't, / Compared with [a being] so inhumane" (6.232–34). Throughout the final scene of the play, he is called a "devil" and a "dog": "Out, dog! Devil!" (214). For Daborne, nothing can redeem the Jew from what he is, not even turning Turk.

Benwash either descends from both Iago and Othello, or he is an analogue to them both; either way, his story makes it all the more likely that Shakespeare acknowledged the Jewish nature of jealousy. At the same time, however, Shakespeare also must have been familiar with the notion that jealousy inhered in the bodies of Moors, a stereotype that no doubt influenced his portrayal of Othello. Cinthio's Disdemona explains her husband's behavior: "Nay, but you Moors are of so hot a *nature*, that every little trifle moves you to anger and revenge"; she concludes that her ill-fated marriage might serve as an example to other Italian ladies "not to wed a man whom *nature* and habitude of life estrange from us."[72] John Leo Africanus notes numerous times in his descriptions of North Africans that they are jealous by nature.[73] As a natural affection or pas-

---

[72] Cinthio, *Moor of Venice*, 21, 28, emphasis added.
[73] Loomba concurs that "Othello ultimately embodies the stereotype of Moorish lust and violence—a jealous, murderous husband of a Christian lady" (*Shakespeare, Race and Colonialism*, 95). Wilson disagrees, arguing that Desdemona correctly sees Othello in terms of an earlier, classical ethnography that saw Africans as "cool" and "dry" by temperament. See Wilson, "Othello's Jealousy," chapter 6 in *English Ethnicity and Race*, 132–60.

sion, jealousy bound Renaissance Jews and Moors in yet another corporeal relation.

The idea of blood relations among Spaniards, Jews, and Muslims was
made official in early modern linguistic usage. John Florio's dictionary,
A worlde of wordes (1598), defined a Marrano as "a Jew, an infidel, a renegado, a nickname for a Spaniard." A later edition is more explicit about
the relationship among these terms; this time, a Marrano is "[a] nickname for Spaniards, that is, one descended of Jews or infidels, and whose
parents were never christened, but for to save their goods will say they
are Christians. Also as Marrana."[74] Shakespeare exposes what he knows
about the composition of Spanish blood when Costard in *Love's Labor's
Lost* calls the Spaniard Armado, "my incony Jew" (3.1.135).

It is essential to note that in his first Venice play Shakespeare had already set up the "relatedness" of Jews and Moors. It should be no sur-
prise, in the context of Renaissance drama, to find that a Prince of Morocco appears in the same play as a Shylock. Yet Shakespeare makes a
point of analogizing them, providing his Moor with sentiments about his
nature, in relation to Christian nature, that approximate the Jew's. In ap-
position to Shylock's "If you prick us, do we not bleed?" the Moorish
Prince demands,

> Bring me the fairest creature northward born,
> Where Phoebus' fire scarce thaws the icicles,
> And let us make incision for your love,
> To prove whose blood is reddest, his or mine.
> (2.1.1–7)

Implicitly, the Prince of Morocco's "incision" recalls his "circumcision"—
both are strikes against his case for a human equality by nature. This
speech was later passed down to Eleazar of *Lust's Dominion*, who makes
the same claim for equality: "Although my flesh be tawny, in my veins /
Runs blood as red, and royal, as the best / And proud'st in Spain."[75] Just
as Shakespeare creates a homology between the Prince of Morocco and
Shylock, I believe he sets up an implicit relation between Othello and
Iago, the "super-subtle Venetian" who pretends to be fighting in the wars

---

[74] John Florio, *A worlde of wordes* (London, 1598); idem, *Queen Anna's new world of
words* (London, 1611), s.v. "Marrano."
[75] *Lust's Dominion*, 1.2, p. 155.

against the general enemy Ottoman, all the while working to compel Othello to "turn Turk."

Shakespeare's association or even confusion of Moors and Turks, typical enough in the Renaissance, has one of its sources in contemporary accounts of the European wars against the Ottoman Empire, in which Moors were often represented as an Ottoman fifth column on Christian soil.[76] Although the Spanish Inquisition may be a source for the inquisitorial paranoia of Shakespeare's Venice plays—their obsession with infidelity—it is the Inquisition of Venice that provides the more immediate source for sixteenth-century conspiracy theories involving imagined racial alliances. Sixteenth-century Venice has been described as a kind of cultural halfway house for Marranos, many of whom emigrated there from Spain and Portugal. While the Jews of Venice were sentenced by law to live in a ghetto (the first Jewish ghetto in Europe, established in 1516), Marranos enjoyed a different and uncertain status: "Poised between the enforcement of strict Catholicism and the granting of liberty of conscience, the city was to become the uneasy refuge of Portuguese New Christians or Marranos, often hesitating between Christianity and Judaism. For some of them Venice would be their point of departure for Ottoman Turkey."[77] After 1550, however, Venice was no longer a refuge for New Christians. Officially there had been an Inquisition in Venice for almost three hundred years, but it gained momentum in the middle of the sixteenth century, during the wars with the Ottoman Empire. Political anxiety was exacerbated by fears that Marranos were using Venice to transport their wealth to Turkey. When war with the Ottomans was renewed in 1570, many Jews were imprisoned as Turkish subjects, and for decades "the maxim that Levantine Jews were all Turkish spies was monotonously repeated."[78] It was, however, the war over Cyprus that brought Venetian anti-Semitism to the fore. In many contemporary accounts, the Turkish invasion was deemed the work of one Joao Miquez Mendes, Duke of Naxos and the Archipelago, the most influential of ex-Christian Jews in Venice. That the Cyprus war was a Jewish war was declared by the Doge in 1570, who said it had been "launched against them

[76] See Andrew C. Hess, "The Moriscos: An Ottoman Fifth Column in Sixteenth-Century Spain," *American Historical Review* 74 (1968): 1–25.

[77] Pullan, *Jews of Europe*, 4.

[78] Randolph C. Head, "Religious Boundaries and the Inquisition in Venice: Trials of Jews and Judaizers, 1548–1580," *Journal of Medieval and Renaissance Studies* 20.2 (1990): 175–204, cites at 180, 184–85.

by the Turk on account of espionage and the evil machinations of Jews."[79]

Although there is no direct proof that Shakespeare knew of the Inquisition of Venice, the "trial" of Shylock is one clue that he knew of the alleged threat that Jews posed there as "aliens" endangering the lives of innocent Christians. And endangered they were: by the time Shakespeare wrote *Othello*, the Turks had long since regained possession of Venice (they regained it, in fact, one year after the victory at Lepanto in 1570). Shakespeare's play seems to be set precisely during the years 1570–71, the same years that Cyprus "underwent a violent conversion from Christian to Turkish rule—the years when it literally 'turned Turk.' "[80] But if anyone is turning Turk in Shakespeare's play, who is to blame, in *Shakespeare's* account of it? Iago.

Among Shakespeare's villains, Iago is most commonly associated with Richard III, Edmund, and Don John. He also resembles Shylock, however, by rhetorical as well as characterological descent. The object of Iago's revenge, like that of Shylock, is to be "even'd" his nemesis' body. Iago's jealousy is a disease—"I'll pour this pestilence into his ear"—a grief that symptomatically results in a desire to grieve Othello, to put Othello into a state of jealousy so strong that "judgment cannot *cure*" (2.1.302). Shylock wants to "feed" his revenge; Iago seeks to "diet" his ("diet" was cited as a chief means of curing physical ailments in Renaissance medical treatises). In a sham oath, pronounced to pique Othello's anguish, Iago calls on heaven, "Good God, the souls of all my tribe defend / From jealousy!" (3.3.175–76); "tribe" is a word that Shylock, and many other Jews and Christians in Renaissance writings, commonly uses to denote a Jewish "kind."[81] Iago is telling the truth, for once, when he tells Othello, regarding his suspicions about Cassio, "I confess it is *my nature's plague* / To spy into abuses, and [oft] my jealousy / Shapes faults

[79] Pullan, *Jews of Europe*, 19, 179.

[80] Emrys Jones, " *Othello*, Lepanto and the Cyprus Wars," *Shakespeare Survey* 21 (1968): 47–52, quote at 52. Jones notes that Cinthio (in his *Moor of Venice*) does not mention any Turkish threat to Cyprus (his story was written long before there was any), so Shakespeare "deliberately brought the action closer to the events of 1570–1" (50).

[81] Shakespeare uses the word "tribe" four times in *The Merchant of Venice*, notably in Shylock's self-identifications as a Jew: "Cursed be my tribe / If I forgive him"; "a wealthy Hebrew of my tribe," "[f]or suff'rance is the badge of all our tribe" (1.3.51–52, 57, 110). Another three uses of the word occur in *Othello*, including Othello's own self-comparison as one who "like the base [Indian / Iudean] threw away a pearl richer than all his tribe" (5.2.347–48).

that are not" (3.3.146–48, emphasis added). "Work on, my medicine"
(4.1.43–44), he incants, as Othello lapses into the "falling sickness" or
epilepsy, a condition associated with Muslims (early modern Christian
accounts of Islam claimed that the prophet Mohammed had the falling
sickness and exploited his illness—perhaps even faked it—to show the
world that he was divinely possessed). Othello's disease is not feigned,
however. His heart is heavy with grief, "Swell, bosom, with thy fraught"
(3.3.449), and turns as hard as Iago's, or almost as hard, as he tells Desdemona before the murder, "Thou dost stone my heart" (5.2.63). "Hard"
or "stony" hearts were long associated with Jews, but also with Turks, as
the Duke insinuates to Shylock, insisting that Antonio's case would even
invoke pity

> From brassy bosoms and rough hearts of flints,
> From stubborn Turks . . . never train'd
> To offices of tender courtesy.
> (4.1.31–33)

Like many stage Jews of the period, Iago is characterized by his desire to
"eat" his enemy's flesh, and he rejoices to see Othello "eaten up with passion" (3.3.391). Poisoned by Iago, Othello is compelled to diet his own
vengeance against his wife and her alleged lovers: "[M]y great revenge /
Had stomach for them all" (5.2.74–75). In his "unkindness," Othello acquires a taste for cannibalism, the crime against (Christian) humanity
long since identified with Jewish barbarity. The "demi-devil" (5.2.301)
Iago is finally condemned, like Benwash and countless other Jews and
Turks on the Renaissance stage as a dog: "O damn'd Iago! O inhuman
dog!" (5.1.63); "O Spartan dog!" (5.2.261).

So too, crucially, is Othello. Famously, Othello takes his own life at the
very moment he recalls the time when, on behalf of the Venetians, he
murdered a Turk: "I took by th'throat the circumcised dog, / And smote
him—thus" (5.2.355–56). It is no coincidence that both Iago and Othello (by his own, final, self-identification) are vilified as "dogs" at the end
of the play. Othello further compares himself to "the base Indian"—or,
alternatively, in a reading I would favor, "the base *Judean*"[82]—who "threw

[82] At this point it will be clear why I prefer the First Folio reading ("Iudean") over the
Quarto reading ("Indian"). For a review of the editorial controversy, see Julia Reinhard Lupton, "Othello Circumcised: Shakespeare and the Pauline Discourse of Nations," *Representations* 57 (Winter 1997): 73–89, quote at 88n25. Lupton argues that "it

a pearl away / Richer than all his *tribe*" (5.2.346–47, emphasis added).
Othello recalls that the Turk he smote was circumcised; the word "thus,"
spoken as he turns the sword of Spain against himself, reminds us that
the Moor is no doubt circumcised as well. Othello, however, is not the
only "dog" in this final scene, as Shakespeare's verbal parallels an-
nounce. Iago too, by implication, is circumcised, weighing in (by Renaissance standards) at at least a pound less than a Christian.

Iago remains alive at the end of the tragedy, leaving the Venetians to
enforce "the censure of this hellish villain, / The time, the place, the torture" (5.2.368–69). Iago had already refused to say the truth about his
motives, but Gratiano promises a forced confession, that "[t]orments
will ope your lips" (306). Lodovico promises more,

> If there be any cunning cruelty
> That can torment him much, and hold him long,
> It shall be his. You shall close prisoner rest,
> Till that the nature of your fault be known
> To the Venetian state.
> (333–37).

Will Iago confess the "nature" of his fault under torture, to the inquisitorial state? Shakespeare himself never confesses it fully, preferring instead to leave us in a state of insecurity about human nature—in all its
(racial) variety. *Othello* does not expose Iago (simply) *as* a Jew, any more
than it conclusively specifies whether Iago "is" a Venetian or a Spaniard.
What is distinctive about Shakespeare's representation of racial nature,
after all, is its hybridity. The point of Othello's final speech, from the perspective of race, is that the Moor positions himself as *both* Venetian and
Turk, not one or the other (not to mention his throwing the "Arabian"
and the "Judean" into the mix).[83] Just as importantly, Shakespeare suggests that the varied and variable racial "parts" of a man may become
proportionally more or less salient under the influence of other men's.
Rather than name a singular identity, Shakespeare defines Iago's nature
according to its context, setting Iago up in the relative position of a Jew

---

was above all the rite of circumcision in its Pauline articulation that emblematized the
affiliation between the Jew and the Muslim in Christian typological thought" (82).
[83] In his final speech, Othello compares his tears to the sap of "Arabian trees"
(5.2.350).

created by so many legendary and literary genealogies of the period, including some of the playwright's own. What matters far more than what
Othello "is" or what Iago "is," for Shakespeare, is what they may become
in relation to others.

In his two Venice plays, Shakespeare considers whether or not "kindness"—in early modern terms, an "equality by nature," and the quality of
mercy or compassion—can or should extend to all the tribes of Renaissance man. The degree to which Shylock, or Iago, or Othello (or Emilia,
or any other character, for that matter) shows us the natural equality that
likens them to others may be exactly the same extent to which it is easiest
to "sympathize" with their pain. The word "sympathy" entered the English language in 1579, and its sources are medical, the "like feeling" engendered within bodies and among them.[84] Galen wrote that parts of the
body that experience pain may not be the site of the disease but a part
acting in sympathy with the real source of illness. Shakespeare seems to
have known this medical concept, as attested by *Othello*, when Desdemona explains, "For let our finger ache, and it endues / Our other
healthful members even to a sense / Of pain" (3.4.146–48).
    In its earliest English usage, however, "sympathy" also referred to an
identification of feeling across bodies rather than within them. The word
gained currency in England after Oswald Croll (ca. 1560–1609) introduced a cure for disease that he called "sympathy ointment." Following
the theories of Paracelsus, Croll believed that a cure for the illness of one
body could be achieved through the "sympathies" between the affected
part of the body and its "signature" in the world. Shakespeare often used
the word "sympathy" in reference to mutual feeling among men or
women: "True sorrow then is feelingly suffic'd / When with like sem-
blance it is sympathiz'd" (*Lucrece*, 1112–13); "O what a sympathy of woe
is this!" (*Titus*, 3.1.148). The early modern semantics of "sympathy" are
analogous to that of "kindness": both make a *material* analogy the pre-
condition for compassion. For Shakespeare's Christian characters, Shylock as a Jew, or Othello as a Moor, are "sympathetic" in proportion to
their physical proximity to Christian feeling and Christian "kindness," to
the extent that they are "equal by nature" to them. It is a rare person, in
Shakespeare's worlds, who can transcend differences of tribal kind, as

---

[84] *Oxford English Dictionary*, s.v. "sympathy."

Desdemona can: "Unkindness may do much, / And his unkindness may defeat my life, / But never taint my love" (4.2.159–60).

Although I have focused in this chapter on gauging some of the racial dimensions of Shakespeare's characters, I must conclude by emphasizing that race is only one, relative part of their compositions and that Shakespeare's characters surely exceed or fall short of any single criterion for evaluation, including nature itself. As a comparison between, for example, Othello and Aaron, of *Titus Andronicus*, immediately reveals, racial natures are not unitary across the plays; if Othello is not a Moor so much as a Moor *of Venice*, the nature of Aaron's Moorishness is conceived in some very unusual juxtaposition with Romans and Goths. Shakespeare recognized potentially shifting degrees of human "equality" and "inequality" among the tribes of man, these relations being no more self-evident to him than any other determination of value discovered through human measurements. By whatever standard we measure a Jew or a Muslim who appears in his plays, Shakespeare insists that we still acknowledge, as even Iago does of Othello, that "another of his *fadom* we have none" (1.2.152, emphasis added).[85]

---

[85] A fadom or fathom is an anthropometric measure; in the Renaissance, it equaled either the length of the forearm or the length of both arms outstretched from fingertip to fingertip (*Oxford English Dictionary*, s.v., "fathom")

SHAKESPEARE'S SOCIAL ARITHMETICS

Checking the Math of King Lear

> There is not, and cannot be number as such. There are several
> number-worlds, as there are several Cultures. We find an Indian,
> an Arabian, a Classical, a Western type of mathematical thought and,
> corresponding with each, an expression of a specific world-feeling . . .
> the soul of that particular Culture.
> —OSWALD SPENGLER, "The Meaning of Numbers"

WESTERN EPISTEMOLOGIES have often assigned mathematics to the domain of objective knowledge, to a realm of laws not of man but of nature or of God. Numbers, of all of our measures of reality, seem to provide the ultimate, abstract, external, impartial, verifiable standard for testing and comparing phenomena, the only one, perhaps, that can be counted on to provide reliable "rules" for assaying our world. Yet the integrity and certitude of mathematical measurement have also been continually contested. What if mathematical laws are not actually "out there" except as projections of the human imagination? What if they are not absolute but subject to variation among cultures or alterations over time? The debate is not a new one:[1] John Dee, in his "Mathematical

---

[1] See Edward MacNeal, *Mathsemantics: Making Numbers Talk Sense* (New York: Viking, 1994); the selections from the history of mathematical literature collected by James R. Newman, *The World of Mathematics* (New York: Simon and Schuster, 1956), especially "Mathematical Truth," "Mathematics as an Art," and "Mathematics as a Culture Clue"; and, most recently, George Lakoff and Rafael E. Nuñez, *Where Mathematics Comes From: How the Embodied Mind Brings Mathematics into Being* (New York: Basic Books, 2000).

Preface" to *Euclid* (as translated by Henry Billingsley in 1570), divided the world into two ontological categories: "Thinges Supernaturall," which are "immaterial, simple, indiuisible, incorruptible, & vnchange-able"; and "Things Naturall," "materiall, compounded, diuisible, cor-ruptible, and chaungeable." Of all we know on earth, only "Thynges Mathematicall" do not fit squarely into either category, "[f]or, these, beyng (in a maner) middle, betwene thinges supernaturall and natu-rall: are not so absolute and excellent, as thinges supernatural: Nor yet so base and grosse, as things naturall." Although Dee conceded that numbers might be "chaungeable," however, he would not allow that mathematics, as a discipline of knowledge, was a human invention: "Nor yet, for all that, in the royal mind of man, [was mathematics] first conceiued."[2] Shakespeare was less sure. At the very least, he acknowl-edged a margin of probable, human error in mathematical operations. This chapter considers Shakespeare's attempts to solve human prob-lems "by the numbers," that is, according to the mathematical measures of what Oswald Spengler might have called the "number world" of Ren-aissance England.

There has been far more research on early modern literacy than on Renaissance numeracy, that is, the ability to think, calculate, and signify in quantitative terms. Although numeracy is not obviously relevant to lit-erary studies, Shakespeare, among many contemporaries, understood mathematics as a signifying system, analogous to other sets of "figures" and forms of language, as his poems and plays attest. "Casting," "ac-counting," "numbering," "figuring," and a host of other terms derived from mathematics are frequently used to describe the "reckoning" of ideas; in fact, such terms are often paralleled with or used metaphori-cally in reference to speaking or writing. Thus when Enobarbus "enu-merates" the affective and discursive forms that fail accurately to gauge Lepidus's love for Mark Antony, he includes counting among them: "Hoo, hearts, tongues, figures, scribes, bards, poets, cannot / Think, speak, cast, write, sing, number, hoo! / His love to Antony" (*Antony and Cleopatra*, 3.1.15–17). In Shakespeare's poems and plays, the terms of mathematics rival language itself as an expressive system. Crucially, numbers provided Shakespeare with a system of signs in which the

---

[2] John Dee, "Mathematical Preface," in *Euclid*, trans. Henry Billingsley (London, 1570), n.p.

meaning of a "figure" depends on its comparison with others; whether or not he might have conceived here an analogy to language, the semiotics of early modern mathematics registers values that are, and can only be, relational.[3]

Shakespeare's numeracy, however, extends beyond a common stock of familiar tropes borrowed from the discipline of practical mathematics. In *King Lear*, especially, mathematics supplied him with a more specialized discourse, and with it a mode of reasoning crucial to his estimates of self and other. In *Lear*, Shakespeare was thinking about how mathematics might be used to calculate just distributions of wealth, a habit of thought that goes back at least as far as Plato. From the start, the play considers the possibilities for "weighing equalities," for approximating quantities of land and goods to human worth and value— value, once again, that can only be determined comparatively.[4] Yet *Lear* also explores the dangers of attempting to correlate numbers to people and numerical equations to relationships among them. While acknowledging its potential utility, Shakespeare suggests that Renaissance mathematics does not provide all the answers and that the human costs of miscalculation may prove immeasurable, more than the numbers can tell.

What's in a number, for Shakespeare? There are many references in Shakespeare's poems and plays to arithmetic, and even to "mathematics," although the word had only been in use since 1581.[5] This shouldn't be surprising. If Shakespeare gained some Latin and less Greek from his purported grammar school education, he must also have had some arithmetic. In the second half of the sixteenth century, first- and second-year English grammar school students spent approximately four hours a week on arithmetic before leaving it behind as a course of study. It is doubtful that Shakespeare had formal training in "higher" mathematics

---

[3] Among Renaissance "number worlds," Pythagorean theories attributed absolute rather than "relative" values to numbers, but as I discuss later in this chapter, I don't believe Shakespeare subscribed to these.

[4] See Barbara Herrnstein Smith, *Contingencies of Value: Alternative Perspectives for Critical Theory* (Cambridge, MA: Harvard University Press, 1988) on the significance of "value" and "evaluation" in literature and culture, a discussion that begins with a reference to *Lear.*

[5] *Oxford English Dictionary*, s.v. "mathematics."

such as algebra,[6] but he seems to have known the basic operations of addition, multiplication, and division, as taught to schoolchildren from the 1540s and through his own time. Shakespeare appears to have acquired more, however, than just the general terms for computing with numbers (of these, "account," "cast," "count," and "figure" occur with special frequency). For example, Shakespeare knew the arithmetic sense of the word "place," first recorded in Robert Recorde's *Ground of Artes* (1543), the standard arithmetic textbook of Shakespeare's day. *The Oxford English Dictionary* cites the *Ground* as the first to include the word to mean, as Recorde defines it, "a seate or roome that a [numerical] figure standeth in."[7] Because of the significance of "place," Recorde explains, the value of numerical figures is permutable: although, for example, the number zero, or the cipher, in Recorde's words, "signifie[s] nothing," it can in the right *place* multiply other figures that go before it by ten, or a thousand, or a million.[8] That's what Shakespeare is talking about in the Prologue to *Henry V*:

> O, pardon! since a crooked figure may
> Attest in little *place* a million,
> And let us, ciphers to this great accompt,
> On your imaginary forces work.
> (15–18, emphasis added)

Or when Polixenes in *The Winter's Tale* explains,

> And therefore, like a cipher
> (Yet standing in rich *place*), I multiply

---

[6] See Geoffrey Howson, *A History of Mathematics Education in England* (Cambridge: Cambridge University Press, 1982), 6–28. I'm doubtful of David Bady's discovery of algebraic equations underlying *The Merchant of Venice*; see Bady, "The Sum of Something: Arithmetic in *The Merchant of Venic*," *Shakespeare Quarterly* 36.1 (1985): 10–30.

[7] *Oxford English Dictionary*, s.v. "place."

[8] Robert Recorde, *Ground of Artes* (London, 1543), 9. See Michele Sharon Jaffee, *The Story of O: Prostitutes and Other Good-for-Nothings in the Renaissance* (Cambridge, MA: Harvard University Press, 1999), for a recent account of the significance of the number zero in Renaissance writings. Jaffe suggests that zero becomes representative of the radically contingent nature of all denotation, because "it means nothing unless joined to others" (40). See also Brian Rotman, *Signifying Nothing* (New York: St. Martin's Press, 1987).

With one "We thank you" many thousands moe
That go before it.
(1.2.6–9, emphasis added)

Shakespeare seems to have been fond of comparing people to "ciphers," as he does in these citations.[9] He is the first writer to use o, the symbol for the arabic zero, metaphorically to mean "[a mere] nothing," as in the Fool's account of Lear: "[T]hou art an o without a figure. I am better than thou art now, I am a Fool, thou art nothing" (1.4.192–94).[10] It is very possible that Recorde suggested this usage to him as well: "There are but tenne figures that are used in Arithmetick; and of those tenne, one doth signifie nothing, which is made like an O, and is privately called a Cypher."[11] For Shakespeare, the value of a man, like that of a number, cannot be known by studying a single figure; as Recorde explains, there is a distinction between a "figure"—a number in isolation—and the value of that figure, which is contingent upon its "place." The mathematical idea of place value is just one way by which Shakespeare understands human identity to be dependent on, and determined by, one's commensuration with others. It may be that in *King Lear*, and perhaps throughout Shakespeare's works, the prospect of a man, a whole life, "signifying nothing" may have been based less on linguistic and more on arithmetic terms of human value.

Shakespeare's use of the word "place" may be the clearest evidence we have of his textbook knowledge of arithmetic, but there are other, corroborating "proofs." Shakespeare refers to the term "parcel" to mean a number that is part of a series of numbers, a usage also first cited in the *Ground of Artes*. When Cleopatra, after Antony's death, feigns humility before Caesar, she (affectedly) mourns "that [her] own servant should / Parcel the sum of [her] disgraces by / Addition of his envy" (5.2.162–64). Although editors often gloss "parcel" as meaning "particularize," it makes more sense to follow Shakespeare's arithmetic rhetoric

---

[9] Other examples include Moth's response to Armado's "A most fine figure!": "To prove you a cipher" (*Love's Labor's Lost*, 1.2.55–56); and Orlando's answer to Jacques's "There I shall see mine own figure": "Which I take to be either a fool or a cipher" (*As You Like It*, 3.2.289–90). The seminal discussion of Shakespeare's preoccupation with "nothing" in *King Lear* is Sigurd Burkhardt's "*King Lear*: The Quality of Nothing," in *Shakespearean Meanings* (Princeton: Princeton University Press, 1968).
[10] *Oxford English Dictionary*, s.v. "o."
[11] Recorde, *Ground of Artes*, 9.

all the way through the line: the "addition" of the servant's envy is part of, or increases, the already considerable "sum" of Cleopatra's shames. In *A Lover's Complaint*, the grieving narrator describes how her seducer overwhelmed her with praises, avouching how "their [the praises'] distract parcels in combined sums" would accrue to her "audit" (230–31). "Parcels," once again, could simply mean "parts" here, but their adding up to the combined sums of an audit suggests, once again, a more technical, arithmetic meaning. Shakespeare's predilection for mathematical language may resolve at least one critical crux: Hamlet, famously, parodies Osric's inflated praise of Laertes through the rhetoric of numbers: "Sir, his definement suffers no perdition in you, though I know to divide him inventorially would dozy th'arithmetic of memory, and yet but yaw neither in respect of his quick sail; but in the verity of extolment, I take him to be a soul of great article" (5.2.112–17). The meaning of the word "article" in this context has always been uncertain, but it seems that Hamlet is responding directly to Osric's promise "to speak sellingly of [Laertes]"—his "quick *sale*"—through continual references to Laertes' "account." Having considered the difficulty of calculating his worth through the "arithmetic of memory" (specifically, by "dividing" him inventorially), Hamlet concludes Laertes' value to be of some undetermined but great "article," which in Recorde's textbook refers to any of the multiples of ten (i.e., ten, twenty, thirty, etc.).

Hamlet's mocking contempt in this passage is typical enough of Renaissance attitudes toward servants with pretensions to "more" than they are, but also toward "arithmeticians"—those who can count and little else. For many of Shakespeare's contemporaries, arithmetic was not the ground of any "higher" art so much as a tool of the trades. Recorde himself had practical, vocational applications in mind for his *Ground of Artes*.[12] "Numbering" is regarded by many of Shakespeare's characters as

---

[12] By 1500 a network of "petty schools" had been established in England to provide basic instruction in arithmetic, in the form of "casting accounts," to apprentices of various trades. There appears to have been a general bias among sixteenth- and early seventeenth-century humanists against providing mathematics education at the university level, although a few championed its merits. Sir John Cheke's advocacy led to university reforms prescribed by the Edwardian Statutes of 1549, according to which all freshmen were required to study math (especially Euclid's geometry) as a foundation for a liberal education. This requirement, however, was removed by new statutes of Elizabeth in 1570, and the study of mathematics was once again generally regarded primarily as preparation for the trades. See Howson, *History of Mathematics Education*, 9–12. Although Recorde begins with a section on what we might call "pure" math, in the form of a dialogue between a teacher and a pupil on the basic operations of

a matter of rote, prosaic, dully methodized, the business of unimagina-
tive or simple-minded boors. Thus Mercutio dismisses Tybalt as "a brag-
gart, a rogue, a villain, that fights by the book of arithmetic" (*Romeo and
Juliet*, 3.1.101–2). Armado, in answer to Moth's multiplication question,
"How many is one thrice told?" replies, "I am ill at reck'ning, it fitteth
the spirit of a tapster" (*Love's Labor's Lost*, 1.2.39–41). Cressida jests that
Pandarus's enumeration of Troilus's "three or four hairs on his chin"
(*Troilus and Cressida*, 1.2.112) won't amount to much praise at all: "Indeed a tapster's arithmetic may soon bring his particulars therein to a
total" (113–14). Iago, meanwhile, expresses disgust at the promotion of
Cassio, the "great arithmetician" (*Othello*, 1.1.19) and "counter-caster"
(31) as he calls him, unworthy to be Othello's lieutenant.

Yet, Shakespeare and his contemporaries were also the heirs of a
philosophical tradition, ultimately attributed to Pythagoras, which identified numbers and numerical relations as the essential, underlying cause
of all phenomena, on earth as in heaven. There was a Judeo-Christian
analogue to this tradition: George Puttenham was among those who
identified human mathematics with the principles of divine creation: "It
is said by such as professe the Mathematicall sciences, that al things stand
by proportion, and that without it nothing could stand to be good or
beautiful. The Doctors of our Theologie to the same effect, but in other
termes say: that God made the world by number, measure, and weight."[13]
As Puttenham observed, this idea was regularly expressed in mathematical treatises of the period, including Dee's "Mathematical Preface": "For
his [God's] Numbryng, then, was his Creatyng of all thinges. . . . And his
Continuall Numbryng, of all thinges, is the Conservation of them in
being."[14] Recorde himself declared that God "wrought the whole world
by number & measure: he is Trinitie in Unity, and Unity in Trinitie," and
chose, as the emblem for his *Ground of Artes*, the words of Solomon's Wisdom: "Thou O God hast ordred all thinges in Measure, Number, and
Weight."[15] Sixteenth- and early seventeenth-century science was driven
by a Christian Pythagoreanism; underlying the investigations of Coperni-

---

arithmetic, a second section is explicitly addressed to merchants and auditors, while a
third, final section treats numbering by the hand for the illiterate.
[13] George Puttenham, *The Arte of English Poesie*, ed. Gladys Doidge Willcock and Alice
Walker (Cambridge: Cambridge University Press, 1970), 64.
[14] Dee, "Mathematical Preface."
[15] Recorde, "Preface to the Lovyng Readers," in *Ground of Artes*.

cus, Kepler, and Galileo was the Pythagorean premise that the universe was ordered by perfect mathematical laws.[16]

If sixteenth-century poets, notably Edmund Spenser, found words to express the spiritual and mystical nature of numbers, as given by the Pythagorean hypothesis,[17] they often had less to say for computing with them. Pythagoras himself was the source of that distinction; he believed that numbers were immaterial, eternal and inviolable, not to be applied as measurements of physical objects. Socrates thus challenged the validity of all computation, asserting, for example, "I cannot satisfy myself that, when one is added to one, the one to which the addition is made becomes two."[18] Even Recorde's "scholar" acknowledges the apparent illogic: "I remember you sayd, I might not adde summes of syndrye thinges togyther, and that I might see by reason."[19] Dee's "Preface" was a notable effort to find a middle way between absolute and relative valuations of numbers.

Shakespeare was not fundamentally interested in numerology, or any absolute rule of mathematics. He concurred with John Selden that "[a]ll those mysterious things they observe in Numbers, come to nothing upon this very ground, because Number is in itself nothing, has not to do with Nature, but is merely of Human Imposition."[20] He also had a far more practical orientation toward numbers than most poets of his age—however much his characters deride "arithmeticians"—as the operational emphasis of his rhetoric suggests (casting, counting, parceling). The early modern art of numbers is especially applicable to *King Lear*, in which the terms of Renaissance arithmetic help determine the outcome of the play. What matters to Shakespeare about mathematics is the possibility of setting up human "equations," with numbers providing comparative measures of men. In *Lear*, above all, Shakespeare reveals the ideological consequences of a long-standing Western rhetorical habit—

---

[16] Morris Kline, *Mathematics in Western Culture* (Oxford: Oxford University Press, 1964), 76–78.

[17] See Alastair Fowler, *Spenser and the Numbers of Time* (New York: Barnes and Noble, 1964). On Shakespeare and numerology, see Alastair Fowler, *Triumphal Forms Structural Patterns in Elizabethan Poetry* (Cambridge: Cambridge University Press, 1970), esp. 183–97.

[18] Quoted in Bady, "Sum of Something," 15.

[19] Recorde, *Ground of Artes*, 43.

[20] John Selden, *The Table-Talk of John Selden* (London: William Pickering, 1847), 133–34.

letting the idea of human "equality" play in the semantic field of mathematics.

> The man that is ignorant of Arythmetyke is nether meete to bee a iudge
> neither a proctour. For how can he well understand an other mans cause
> appertainng to distribution of goodes.
> —RECORDE, *Ground of Artes*

King Lear begins, famously, with a problem in division and "distribution of goodes." It opens in the middle of a conversation between Gloucester and Kent about a confusion regarding the King's plans:

KENT: I thought the King had more affected the Duke of Albany than Cornwall.
GLOU: It did always seem so to us; but now in the division of the kingdom, it appears not which of the Dukes he values most, for [equalities] are so weigh'd, that curiosity in neither can make choice of either's moi'ty.
(1.1.1–7)

The confusion concerns which of his sons-in-law Lear "values most." Everyone seems to think that it was always Albany, but now there's no way to tell: "equalities are so weigh'd" in the division of the kingdom that the King would appear to like them equally well.[21] The division of the kingdom, from the start, is presented as the act of approximating material with human value, to make land and goods proportionate with love and affection. Turning to Edmund, Kent and Gloucester apparently change the subject and never return to it. Yet in terms of the business of human measurement, at least, they stick to their topic—judging what people are worth and what they should get in return for it. Gloucester tells Kent that Edgar, though his only legitimate heir, is yet "no dearer in [his] account" (20–21) than his bastard son Edmund; equalities are so weighed in his love for his two children, he might have said, that neither one can be

---

[21] Two quarto editions have "equalities are so weigh'd" where the First Folio gives "qualities are weigh'd"; either way, Gloucester's point is that the dukes appear to have been apportioned the same shares. The choice of "qualities" perhaps stresses the idea that the two shares are different, though valued as "equal."

counted the "dearer." Yet even as he offers equal shares of his affection
for his two sons, Gloucester whisks the bastard away; no wonder that Edmund, from his entrance in the play, promises "to study deserving" (31)
further. Studying what he deserves, what he's worth relative to the worth
of his legitimate brother, is indeed Edmund's chief discipline throughout
the play.

The opening dialogue between Gloucester and Kent also reveals that
the King has, in advance of the "love-test," decided on the relative value
of his children and divided his kingdom accordingly (although it is also
possible that Lear does indeed value Albany more than Cornwall, just as
is presumed, but has decided to give them equal shares nonetheless). In
either case, Lear never attempts to hide the fact that he has already made
his cuts:

> Give me the map there. Know that we have divided
> In three our kingdom.[22]
>
> . . . .
>
> We have this hour a constant will to publish
> Our daughters' several dowers.
> (37–38, 43–44).

He makes it clear that one of the dowers is greater than the others:
"Which of you shall we say doth love us *most*, / That we our *largest* bounty
may extend / Where nature doth with merit challenge?" (51–53, emphasis mine). Although Regan adds to Goneril's self-assessment ("I am made
of that self metal as my sister, / And prize me at her worth" [69–70]) by
professing an even greater love for her father, she receives exactly the
same share, just as Lear had predetermined:

> To thee and thine hereditary ever
> Remain this ample third of our fair kingdom
> No less in space, validity and pleasure,
> Than that conferr'd on Goneril.
> (79–82)

---

[22] Note that the "measuring of ground by length and breadth" along with the weighing of coin, are the chief applications of arithmetic, according to Recorde ("Preface
to the moste Christian Prince Edward the Syxthe," in *Ground of Artes*).

Presumably, it's no *more* in space, validity, or pleasure, either. Lear then explicitly directs Cordelia to say something that will win her the "third more opulent" (86) than those granted to her sisters.[23] Cordelia, however, assesses her own love as too "ponderous" to be "heaved" up from within (78, 91). Indeed, her valuation of her own self-worth finally outweighs that of her material worth, her dowry: "Then poor Cordelia! / And yet not so" (77–78).

It has been suggested that the attempt to quantify love in act 1, scene 1, is characteristic of the moral order of Goneril and Regan, who, as William Elton puts it, "move within a universe of confused proportions in which the only unit of measurement is quantitative, and the main value word, 'more' "; the "ethic of calculation," he argues, applies only to the villains of the play.[24] For Elton, Cordelia, in contrast, seems to know, with Antony and Cleopatra, that "[t]here's beggary in the love that can be reckoned" (1.1.14–15), that her love is too "ponderous," too weighty, to be measured in words. Yet although she rejects the "glib and oily art" (1.1.224) of exaggerated praise, Cordelia is not averse to the "ethic of calculation" altogether, the idea that one's love should be measured and divided in "right fit" ways (97). Cordelia thus asserts a due proportion of love for the King, "no more, no less," (93) and promises to give "half" (102) her love to her husband, in contrast to her sisters who claim to love their father all. *King Lear* does not unequivocally refute arithmetic and the effort to evaluate human worth or to credit it through material rewards. When Albany, at the end of the play, seeks to mete out justice, he replays the opening scene by granting final dispensation to the survivors:

> You, to your rights,
>   With boot, and such addition as your honors
>   Have more than merited. All friends shall taste
>   The wages of their virtue.
> (5.3.301–4)

"Boot" and "wages" no doubt refer here as much to material and other worldly compensation as to Albany's heartfelt thanks. Albany implies,

---

[23] Lear's arithmetic may be faulty here as well, if "third" means one of three *equal* parts; by this denotation, his "third more opulent" is more bad math. But the word "third" might simply mean in this context "the last of a successive group of three." See the *Oxford English Dictionary*, s.v. "third."
[24] William Elton, *King Lear and the Gods* (Lexington: University Press of Kentucky, 1988), 121–22.

moreover, that some merit "more" wages than others, some less, in proportion to their "honors." Some "addition," some "more," conceived in mathematical terms, *is* due to those worthy of it. For Cordelia, as for Albany, it's a matter of trying to get the math right.

It's not just Edmund but *King Lear* as a whole that studies deserving. The play is framed by scenes of distributive justice or "weighing equalities," the division of goods according to personal deserts.[25] Theories of "division" in the play do not derive solely from contemporary works on arithmetic, however, but ultimately from works on social and political theory, notably Aristotle's *Nichomachean Ethics*, perhaps via Thomas Elyot's *Book Named the Governor* (1531); these works established the rhetorical link between mathematical and social thought. It is Aristotle and his English exponents who provided Shakespeare with a model for a Western "social arithmetics."[26] Crucially, for Aristotle, all forms of justice are based on "equality": "[S]ince an unjust man is the one who is unfair, and the unjust is the unequal, it is clear that corresponding to the unequal there is a mean, namely that which is equal. . . . If then the unjust is the unequal, the just is the equal—a view that commends itself to all without proof."[27]

Joel Kaye has recently revisited Aristotle's *Ethics* and concludes that "every detail of his discussion of justice reveals that he was conceptualizing and treating the world as if real, existent problems or qualities (justice, service, virtue, gain, loss) could be represented by lines or numbers and solved mathematically"; moreover, all mathematical thinking in Aristotle represents a search for "equality" in the human sphere.[28] Thomas Aquinas concurred: "[J]ustice denotes a kind of equality—as the name

---

[25] Elton also discusses the theme of distributive justice, including Lear's original "maldistribution," arguing that the play lays bare the "marked disproportion in the workings of human justice" and in cosmic justice as well. See Elton, *King Lear and the Gods*, especially 225–28.

[26] My phrase "social arithmetics" is derived from Jeremy Bentham's "moral arithmetics," but the idea of applying mathematical models to matters of social ethics is an ancient one.

[27] Aristotle, *Nichomachean Ethics*, trans. H. Rackham, Loeb Classical Library (London: William Heinemann, 1947), V.3.1–3. See Elaine Scarry, *On Beauty and Being Just* (Princeton: Princeton University Press, 1999), for a discussion of the significance of the ideas of "equality" and "balance" in the history of aesthetics and the history of social theory.

[28] Joel Kaye, *Economy and Nature in the Fourteenth Century: Money, Market Exchange, and the Emergence of Scientific Thought* (Cambridge: Cambridge University Press, 1998), 42–43.

itself shows, for it is commonly said that things are 'adjusted' to one another when they are made equal, and equality has to do with the relation
of one man to another. . . . The 'right' or 'just' is defined by the commensuration of one person with another. . . . For nothing is equal to itself, but only to something else."[29] Aristotle's theory of justice as equality
was reiterated by Elyot along with many other Renaissance English writers. Thomas Wilson, for example, in his *Discourse upon Usury* (1572), concurs that "Iusteice is none other thinge then a certeine evenhode or
equalitie, and therefore they that do not in their dealings use an equal
property, do not use Iustice."[30] But what does "equality" mean in these
contexts? "Equal" in early modern English may denote sameness or
"identity" of amount, quality, or degree (as it does today), but not always.
In one derivation from Greek usage, for example, "equality" can mean
"evenness" or moderation, neither too much nor too little, an unspeci-
fied quantity that represents the mean or middle ground between extreme terms. As a mathematical term, the English word "equal" dates
from Billingsley's translation of *Euclid* and "equation" from Dee's "Mathematical Preface" to that work.[31]

That "equal" does not always mean the "same" in early modern contexts is perhaps most clearly evidenced in Spenser's *Faerie Queene*, especially as it is used in book 5, "Of Iustice." Artegal is brought up by the
goddess Astraea "to weigh both right and wrong / In equall balance";
"equality" is clearly the basis of justice for Spenser, as it was for Aristotle:
"For equall right in equall things doth stand." Yet when the Gyant of
book 5, canto 2, attempts to redistribute every "surplus" so that both
sides of every natural and social scale are reduced to an "equality," Artegal condemns him: "Thou that presum'st to weigh the world anew, / And
all things to an equall to restore, / In stead of right me seemes great
wrong dost shew."[32] "Equality" does not mean the same thing for the
giant as it does for Artegal, nor does it "add up" to the same number.[33]

[29] Thomas Aquinas, *Summa theologiae*, in *St. Thomas Aquinas: Political Writings*, ed. and
trans. R. W. Dyson (Cambridge: Cambridge University Press, 2002), IIaeIIae 57, pp.
159, 166; IIaeIIae 58, p. 172.
[30] Thomas Wilson, *A Discourse upon Usury* (G. Bell, 1925), 287.
[31] *Oxford English Dictionary*, s.v. "equal" and "equation."
[32] Edmund Spenser, *The Faerie Queene*, ed. Thomas P. Roche Jr. (New York: Penguin,
1978), 5.1.7.1–2, 5.4.19.1, 5.2.34.1–3.
[33] See chapter 5 for further discussion of this episode and Shakespeare's account of
the legal consequences of the idea of judgment as "equality."

The fact is, it is difficult to define Renaissance "equality," what it means, precisely, as an outcome of early modern socioeconomic measurements. That's because Spenser, Shakespeare, and their contemporaries inherited a problem with equality that was, ultimately, a rhetorical one. Plato wrote,

> There are, in fact, two equalities under one name but, for the most part, with contrary results. The one equality, that of number, weight, and measure, any society and any legislator can readily recure in the award of distinctions, by simply regulating their distribution by the lost, but the true and best equality is hardly so patent to every vision. . . . For it assigns more to the greater and less to the lesser, adapting its gifts to the real character of either. In this matter of honors, in particular, it deals proportionately with either party, ever awarding a greater share to those of greater worth. . . . [J]ustice we explained to be a true and real equality, meted out to various unequals.[34]

For Plato, "equality" is not a fact about human nature but rather an act of human measurement, or rather two different acts of human measurement "with contrary results." So contrary in fact that one seems to treat all men as the same and the other as if they were "greater" or "lesser" than one another. There are according to Plato *two* kinds of equality, and one nearly contradicts the other. Why, then, do they share the same name?

Aristotle and his Renaissance heirs made clear why: "equality" was ultimately a mathematical concept, and mathematics resolved the apparent rhetorical inconsistency in the meanings of the word. In Elyot's paraphrase of Aristotle, there are two kinds of mathematical equations, which in turn provide the basis for two kinds of human measurement and human justice. First, "arithmetic proportion," he explains, is a simple one-to-one equivalence; in the human sphere, arithmetic proportion rules matters of "buying and selling, loan, surety, letting, and taking, and all other thing wherein is mutual consent at the beginning." The arithmetic proportion between men de-

---

[34] Plato, *Laws*, trans. R. G. Bury, Loeb Classical Library (Cambridge, MA: Harvard University Press, 1926), VI.757.

mands what he calls "corrective justice," which "hathe no regarde to the persone, but onely considerynge the inequalitie whereby one thynge excedeth the other / indeuoreth to brynge them bothe to an equalitie."[35] "Distributive justice," on the other hand, is based on "geometric" equality; by "geometric" Aristotle was referring to proportional equivalences or ratios. As he explains it, "[T]his kind of proportion is termed by mathematicians geometrical proportion; for a geometrical proportion is one in which the sum of the first and third terms will bear the same ratio to the sum of the second and fourth as one term of either pair bears to the other term."[36] In contrast to corrective justice, distributive justice has regard to the person and adheres to the principle "to euery manne his owne."[37] It follows, then, that "if the persons are not equal, they will not have equal shares." Aristotle observes that "it is when equals possess or are allotted unequal shares, or persons not equal equal shares, that quarrels and complaints arise." By Aristotle's calculations, both arithmetic and geometric "equality" are examples of a mathematically deducible "mean," as he explains it:

> Now of everything that is continuous and divisible, it is possible to take the larger part, or the smaller part, or an equal part, and these parts may be larger, smaller, and equal either with respect to the thing itself or relatively to us; the equal part being a mean between excess and deficiency. By the mean of the thing I denote a point equally distant from either extreme, which is one and the same for everybody; by the mean relative to us, that amount which is neither too much nor too little, and this is not one and the same for everybody. For example, let 10 be many and 2 few; the one takes the mean with respect to the thing if one takes 6; since $6 - 2 = 10 - 6$, and this is the mean according to arithmetical proportion. But we cannot arrive by this method at the mean relative to us. Suppose that 10 lb. of food is a large ration for anybody and 2 lb. a small one: it does not follow that a trainer will prescribe 6 lb.,

---

[35] Sir Thomas Elyot, *The Booke Named the Governor*, facsimile ed. (Menston: Scolar Press, 1970), 171.

[36] Aristotle, *Nichomachean Ethics*, trans. and ed. Roger Crisp (Cambridge: Cambridge University Press, 2000), V.3.13.

[37] Elyot, *Booke Named the Governor*, 172.

for perhaps even this will be a large ration, or a small one, for the particular athlete who is to receive it.[38]

According to Aristotle, there are two kinds of "means" just as there are two kinds of equalities, one "according to arithmetic progression," which does not take individual differences into account, and another which depends on individual needs and merits, on "every man [receiving] his own."

Aristotle admits that it's a lot easier to agree on arithmetic equalities or means than geometric ones, because "[a]ll are agreed that justice in distributions must be based on desert of some sort, although they do not all mean the same sort of desert." For example, in the political sphere, "democrats make the criterion free birth; those of oligarchical sympathies wealth, or in other cases birth; upholders of aristocracy make it virtue." It depends on what one chooses to "count" about people. Yet however one chose, the basis of human justice in number offered grounds to argue that certain social practices (including, for Aristotle, slavery) were based on "equality": "[P]olitical justice . . . is found among people who associate in life to achieve self-sufficiency, people who are free and either proportionally or arithmetically equal."[39] "*Either* proportionally or arithmetically*": both were demonstrable, mathematical "equalities." That Aristotle's math was still operative as the ground of Renaissance arts of law is amply evident. Recorde writes, "In Lawe twoe kyndes of Justice are the somme of the studie: Iustice Distributive, and Iustice Commutative. . . . But what is any of the bothe without Nomber?"[40] Francis Bacon in his *Advancement of Learning* (1605) asks rhetorically, "[I]s there not a true coincidence between commutative and distributive justice, and arithmetical and geometrical proportion?"[41] In the Renaissance, the rhetoric of mathematics not only subsumed different accounts of justice under the same terms but provided proof that they all remained deducible by mathematical "equations," all

---

[38] Aristotle, *Nichomachean Ethics*, trans. H. Rackham, Loeb Classical Library (London: William Heinemann, 1947), II.6.4–7.

[39] Ibid., V.3.7–8, V.6.1134a.

[40] Robert Recorde, *Whetstone of Witte* (London, 1557), sig. B1v.

[41] Francis Bacon, *The Advancement of Learning*, in *The Works of Francis Bacon*, vol. 5, ed. James Spedding, Robert Leslie Ellis, and Douglas Denon Heath (New York: Garret Press, 1968), 348.

worked out to the same social result, despite the difference in the numbers attained.

Understanding the early modern rhetoric of mathematics, as applied to people, is essential to resolving what appears to be the uneven politics of *King Lear*—a play that for many readers seems to invoke the most radical statements of egalitarianism in Shakespeare's canon, but without making any serious recommendations for social change. For some, *King Lear* is representative of the kind of contained subversion critics have frequently discerned in the politics of Shakespeare's drama.[42] However, the revelations about equality as experienced collectively by those exiled onto the heath are in fact applied consistently to the play's conclusion— or rather, consistently in terms of the system for evaluating social relationships reworked here by the playwright.

King Lear dramatizes Recorde's idea that arithmetic is the ground of the arts of justice: from the start, Lear attempts to weigh "equalities" among his children, in all the equivocal early modern senses of the word. It's just that his math (which is the same here as saying his judgment) is not very good. He not only makes the fateful error regarding Cordelia's true value but had already miscalculated by giving "equalities" to Albany and Cornwall (a more discerning judge might have given Albany more). Even worse, he then divides Cordelia's "third more opulent" share, leaving his sons-in-law with equal halves of the kingdom. He further miscalculates the solution to his own problem of dividing up the kingdom, that is, what will be left to him as "remainders"—another arithmetic term (1.4.250).[43] Lear divides up everything—even the crown itself—and yet claims he will still retain "the name, and all th'*addition* to a king" (1.1.136, emphasis added). Lear's preoccupation with retaining the "name" of king is familiar enough among Shakespeare's royals, notably Richard II, who also has a tendency to overestimate the value of his title in and of itself. Although "addition" is generally glossed, appropriately enough, as referring to further distinctions or honors, the mathematical sense of the word is no coincidence here. In arithmetic terms, how can one divide up everything and still have something added, something

---

[42] Jonathan Dollimore, for example, in *Radical Tragedy: Religion, Ideology and Power in the Drama of Shakespeare and His Contemporaries* (Chicago: Chicago University Press, 1984), has shown how various "revolutionary (emergent) insight[s] are folded back into a dominant ideology" in *Lear* (201).

[43] The word "remainder" appears in Recorde's *Ground of Artes*, 39.

"more"?[44] It's no wonder that the Fool continually challenges the king's command of numbers:

> FOOL: The reason why the seven stars are no moe than seven is a pretty reason.
> LEAR: Because they are not eight.
> FOOL: Yes indeed, thou wouldst make a good Fool.
> (1.5.34–38)

He tests him with a mathematical riddle:

> Have more than thou showest,
> Speak less than thou knowest
> Lend less than thou owest
> Ride more than thou goest,
> Learn more than thou trowest,
> Set less than thou throwest;
>
> .   .   .   .
>
> And thou shalt have more
> Than two tens to a score.
> (1.4.118–23, 126–27)

The Fool's advice takes the form of a series of additions and subtractions that seems to leave the beneficiary of his wisdom coming out ahead ("And thou shalt have more"), but that "more" is only apparent; for once, Lear reckons correctly when he deduces that the difference between "two tens" and a "score" is nil: "This is nothing, Fool" (128). But it is too little, too late. Lear's miscalculations of his daughters' "merit" bring down a kingdom, and for Shakespeare it is doubtful that there is any mathematical solution to the nullity that results.

Goneril and Regan, meanwhile, attempt to devalue their father by reducing and finally annulling the king's company of retainers. Reiterating the bad math according to which he claimed to divide all yet still maintain an "addition," Lear reminds them,

---

[44] Richard Halpern sees Lear's faulty expectation of retaining some "addition" as his failure to understand the zero-sum economy that now defines his political and social world; Halpern, *The Poetics of Primitive Accumulation: English Renaissance Culture and the Genealogy of Capital* (Ithaca: Cornell University Press, 1991), 253.

I gave you all—

.   .   .   .   .   .

But kept a reservation to be followed
With such a number
(2.4.249, 252–53)

Lear attempts to retain as many followers as he can as the sisters, in
Goneril's terms, "disquantity" (1.4.249) his train, while Lear reminds
them of the presumed equation between numerical and affective value:
"Thy fifty yet doth double five and twenty, / And thou art twice her love"
(2.4.258–59). When Regan makes it clear that this will be, for Lear, another arithmetic reduction to zero ("What need one?" [263]), Lear cries
out, in his famous speech,

O, reason not the need! our basest beggars
Are in the poorest thing superfluous.
Allow not nature more than nature needs,
Man's life is cheap as beast's.
(2.4.264–67)

Lear's objection to their "reasoning"—in early modern mathematical
treatises, defined as the capacity to compute with numbers[45]—may seem
irrational, out of all proportion, just as his tirade over Cordelia's earlier
recalcitrance did; losing his band of retainers may not seem loss enough
to provoke a consideration of the value of human life. Yet in the context
of *King Lear*, it is a "reasoned" (if impassioned) response. Lear is not re-
jecting the attempt to calculate human need, to reduce it to something
quantifiable, any more than Cordelia treats love as "immeasurable" in
the first scene of the play. It's not that he thinks "need" can't be reckoned

---

[45] Many early modern arithmeticians defined "reason" as the capacity to compute
with numbers. Thomas Masterson, *First booke of arithmeticke* (London, 1592), described
reason as "the habitude, regard, respect, or comparison, that two quantities of one
kind haue together, according to quantity" (2). Following the idea that God's creation
depends on number and that the ability to compute is evidence of our link to the divine, many suggest, with Thomas Hill, that "numbring [is] (almost) al, between a
man and beast"; Hill, *The arte of vulgare arithmeticke* (London, 1600). Later in the century, Thomas Hobbes famously identified "reason" with mathematical "reckoning."
See Lakoff and Nunez, *Where Mathematics Comes From*, 357, for a brief history of this
identification, culminating in the rise of symbolic logic as an analogue to the rational
faculty.

mathematically (need or *indigentia* was a well-established criterion for measuring economic value);[46] on the contrary, he says it's quite possible to count a "superfluity" of things even among those who have nearly nothing at all. Lear is not objecting to the quantification of human need but rather to the idea that "need" is the appropriate criterion for judging what is enough for him. After all, Lear's proposal, in act 1, scene 1, was to extend his largest bounty "where nature doth with merit challenge"—implying that "merit" competes with nature as a criterion of worth. Only to meet the bare exigencies of survival in nature, Lear suggests, is to live a life "*cheap* as beasts"; to have human value means having more than nature needs, some social superfluity, some "addition." Without that "addition," he figures, we are "nothing." I will return to the significance of what must be "added" to nature, the "more" that makes us distinctively human, later in this chapter.

Lear's encounter with Edgar, disguised as Tom O'Bedlam, leads the king to his most explicit accounting of human justice as a problem of social arithmetics, the proper apportionment of goods to men. "Poor Tom" embodies Lear's vision of a life as cheap as a beast's: "[U]naccommodated man is no more but such a poor, bare, fork'd animal as thou art" (3.4.106–8). Edgar had already described his adopted guise as "the basest and most poorest shape / That ever penury, in contempt of man / Brought near to beast" (2.3.7–9). *King Lear* suggests that poverty cheapens, debases the value of a man's life, and in ways that can be and must be accounted. One man's "too little" is referred, relatively, to another's "too much":

> Poor naked wretches, wheresoe'er you are,
> That bide the pelting of this pitiless storm,
> How shall your houseless heads and unfed sides,

---

[46] Kaye, *Economy and Nature*, reminds us that Aristotle marked human need as the standard against which all economic value is determined, and his commentators, including Thomas Aquinas, concurred: "The one thing that measures all things . . . is need [*indigentia*]. . . . In exchange, things are not valued according to the dignity of their natures [*dignitatem naturae*]. . . . But the price of things is determined according to how much men need them because of their usefulness" (70). At least one late medieval economic thinker, Jean Buridan, argued that even need itself might be relative so that the rich "needed" different things from the poor; on these grounds he suggested that an analysis of economic value had to include the worth not only of basic goods but also of luxuries like jewels and fine clothes (Kaye, *Economy and Nature*, 147).

Your [loop'd] and window'd raggedness, defend you
From seasons such as these? O, I have ta'en
Too little care of this! Take physic, pomp,
Expose thyself to feel what wretches feel,
That thou mayst shake the superflux to them,
And show the heavens more just.
(3.4.28–36)

As has been often noted, Gloucester appears to arrive at the same conclusions about the poor. Handing "Tom" his purse, Gloucester cries out to the heavens:

Let the superfluous and lust-dieted man,
That slaves your ordinance, that will not see
Because he does not feel, feel your pow'r quickly;
So distribution should undo excess,
And each man have enough.
(4.1.67–71)

The matter once again is distributive justice, the proper apportionment of goods to men. The play's apparent demand for a reallocation of superfluous material comforts to the poor has been variously interpreted, in ideological terms: some see here a traditional Christian concern for almsgiving, while others hear a proto-Marxist call for the redistribution of wealth.[47] Although it has been shown that words such as "distribution," "superfluity," and "excess" were common in Renaissance treatises on charity, they are also characteristic of the Aristotelian discourse on ethics, justice, and the "mean." It seems more consistent with the play's pre-Christian (or post-Christian) ethos to assume that Lear and Gloucester are once again referring to the problem of "division" as a factor of Aristotelian justice. In *Lear*, "enough" is not a fixed sum, an "equal" or identical quantity derived arithmetically; its determination depends on a fuller, comparative, relative evaluation of self and other.

Self-assessment and the assessment of others are not two different operations in *King Lear*. Lear's judgment of his daughters at the start of the

---

[47] See Judith Kronenfeld's review of recent approaches to the question of "distribution" in Lear, in *King Lear and the Naked Truth: Rethinking the Language of Religion and Resistance* (Durham, NC: Duke University Press, 1998), 170–99.

play is flawed, in just proportion to the fact that "hath ever but slenderly known himself." In his section on justice, Elyot warns,

> If thou be a gouernour / or haste ouer other soueraygntie / know thy selfe. That is to saye / knowe that thou art verely a man compacte of soule and body / and in that all other men be equall vnto the. Also that euery man taketh with the / equall benefite of the spirite of life / nor thou haste any more of the dewe of hevyn / or the brightnes of the sonne / than any other persone . . . and [remember] that your body is subiecte to corruption / as all other be / & life tyme vncertayne.[48]

The Renaissance invocation of the Socratic injunction to "know thyself" was rarely if ever meant to suggest that one ought to contemplate one's distinctive, individuated identity.[49] The charge was rather to know what one shares with others, what it is to be a man as other men. Stripped of the additions of court and society, Lear discovers "feeling," the sensations of his mind and his body. Lear recognizes that in having all and being all he had believed himself invulnerable to such feelings, but "'Tis a lie, I am not ague-proof" (4.6.105). Lear's new sense of his identity correlates precisely with his identification with "poor Tom," and from there, with all "poor, naked wretches." To know oneself, as Elyot puts it, is to understand one's *equalities* with others: "Than in knowinge the condicion of his soule & body / he knoweth him selfe / and consequently in *the same thinge* he knoweth *euery other man*." Moreover, as Elyot writes,

> Thy dignitie or autorite / wherin thou onely differest from other is (as it were) but a weighty or hevy cloke / fresshely gliteringe in the eyen of them that be poreblynde. . . . [I]t may be shortely taken of him that dyd put it on the / if thou vse it negligently: or that thou weare it nat comely and as it appertaineth. Therfore whiles thou wearest it / know thy selfe / knowe that the name of a soueraigne or ruler without actuall gouernaunce is but a shadowe / that gouernaunce standeth nat by wordes onely / but principally by acte and example.[50]

---

[48] Elyot, *Booke Named the Governor*, 176–78.
[49] Anne Ferry, *The Inward Language: Sonnets of Wyatt, Sidney, Shakespeare, Donne* (Chicago: Chicago University Press, 1983), 40.
[50] Elyot, *Booke Named the Governor*, 177 (emphasis mine).

There is much here that Shakespeare seems to have borrowed directly—the idea, again, of a human "equality by nature"; the likening of social and political authority to a cloak, which glitters in the eyes of the "pore-blynde"; and the idea of the "shadow" king, ruling in word or name alone. Lear's evolving understanding of a common human nature leads him to strip himself of the final shreds of his clothing, to subtract the additions of society to reveal another "poor, bare, fork'd animal": "Off, off you lendings! Come, unbutton here" (3.4.108–9). In his last words he expresses the same desire: "Pray you undo this button" (5.3.310). Fellow feeling or compassion marks the beginnings of a new society in *Lear*, one to be based on shared experience ("speak[ing] what we feel" [5.3.325]) and led by a man "who, by the art of known and feeling sorrows, / [Is] pregnant to good pity" (4.6.222–23). As Edgar, still disguised as poor Tom, considers, "[T]he mind much sufferance doth o'erskip, / When grief hath mates, and bearing fellowship" (3.6.106–7).

By Renaissance standards, however, equality by nature is not "enough" to sustain human fellowship in the world of *King Lear*, even if it provides its basis. *King Lear* does not, in the final analysis, advance the creation of a commonwealth of equals based solely on the substance of the body and soul. It is Edmund, above all, who dramatizes the dangers of measuring human value and human deserving by natural equalities alone. The bastard begins with a proper assessment of what he shares with his legitimate brother:

> [M]y dimensions are as well compact,
> My mind as generous, and my shape as true,
> As honest madam's issue.
> (1.2.7–9)

Edmund asserts that he is "created equal," in mind and in body, to other men. Edmund's criterion of measurement, equality by nature, becomes the basis of his theory of justice and social revolution: "This seems a fair deserving, and must draw me / That which my father loses: no less than all" (3.3.23–24). It may seem contradictory in the context of the tradition of liberal humanism, but by Renaissance measures, Edmund's theory of "natural equality" leads directly to inequality, not to his drawing the "same" as others but rather, potentially, "no less than all." Elyot had warned of the political consequences of Edmund's ideas of equality:

> [H]owe farre out of reason shall we iudge them to be / that wolde exterminate all superioritie / extincte all gouernaunce and lawes / and vnder the coloure of holy scripture / whiche they do violently wraste to their purpose / do endeavour them selves to bryng the life of man in to a confusion ineuitable / & to be in moche wars a state than the afore named beestes. . . . Than were all our equalitie dasshed.[51]

Crucially, Edmund's theory of equality, for Elyot (as for Aristotle) ultimately subverts equality: although the just polity is based on a presumed natural equality, at the same time that notion, if used as an *exclusive* measure of human value, threatens to undermine the very equality on which the individual and the state must rest. Put another way, it is not only measuring how men and women are the same but how they are different, not only how they are equal but the ways in which they are unequal, that must determine who they are and what they correspondingly deserve. As Goneril puts it, "O, the difference of man and man!" (4.2.26). In the mathematical terms set by the play, however, paradoxical as it may seem, it adds up to the same thing—an "equal" society.

Commensurate feeling may be a reliable "proof" of equality by nature in *King Lear*, but how does one gauge merit? As opposed to what makes us equal, which can be discovered in the material body, what makes us unequal in *Lear* is "hardly so patent," as Plato put it. If Elyot asserts the natural equalities shared by all men, he also insists on human "qualities" differentiated by degree:

> In semblable maner the inferior persone or subiecte aught to consider / that all be it (as I haue spoken) he in the substaunce of soule and body be equall with his superiors yet for als moche as the powars and qualities of the soule and body / with the disposition of reason / be nat in euery man equall / therefore God ordayned a diuersitie or preeminence in degrees to be amonge men / for the necessary derection and preseruation of them in conformitie of lyuinge.[52]

---

[51] Ibid., 179.
[52] Ibid., 178–90.

Lear's proposal, in act 1, scene 1, is to extend his largest bounty "where nature doth with merit challenge" (53), implying that "merit" (whatever it is and wherever it lies) cannot be referred solely to the substance of "nature." Merit represents some undefined "preeminence" or "addition" to our *common* nature; what Lear and Gloucester discover is that some "addition," something more, is necessary for human life everywhere: "I will piece out the comfort with what addition I can" (3.6.2–3). In the signifying system of *King Lear*, "addition" stands for the social accommodations of nature—clothes, honors, degrees—but also, it seems, rational and moral "powers" in excess of our commonalities, our shared feelings, and our mortality. These additions represent that which distinguishes us from other people, the diversity of our (e)qualities. In the social arithmetics of Shakespeare's play, we are "nothing"—at least, nothing human—without them.

The King of France, in *All's Well That Ends Well*, makes the same distinction between our natural equalities and our "superadded" ones. In his effort to convince the recalcitrant gentleman Bertram to marry a physician's daughter, Helen, he observes:

'Tis only title thou disdain'st in her, the which
I can build up. Strange it is that our bloods,
Of color, weight, and heat, pour'd all together,
Would quite confound distinction, yet stands off
In differences so mighty.
(2.3.117–21)

Yet as king, he explains, he can level their remaining differences, no matter how "mighty":

If thou canst like this creature as a maid,
I can create the rest.
 .   .   .   .
We poising us in her defective scale,
Shall weigh thee to the beam.
(142–43, 154–55)

Fulfilling his promise to Helen to "make it even" (2.1.192) between them, the King mends her "defective scale" with "great additions" (2.3.127) of his own making, leveling the social and economic inequalities that divide Helen from her beloved. In the fairy-tale economy of *All's*

Well, Shakespeare suggests that human additions are all "created" by men rather than "given" at birth; they belong to the realm of art rather than of nature.

Lear's faithful servant Kent, like Helen of *All's Well*, is offered a material "addition" in return for his moral deserts. What is to be the fate of Bedlam beggars in the final socioeconomic dispensation of the play? Although Gloucester gives "poor Tom" his purse, even Kent "shun[s] [his] abhorr'd society" (5.3.211), and it seems doubtful that "houseless poverty," in general, is included in Albany's final promise of compensation to the deserving.[53] How then do the social arithmetics of *King Lear* account for this disparity?

There is a mathematical solution or at least an early modern mathematical solution to the problem. If "unaccommodated man" is, as Lear puts it, "the thing itself," he is also "no thing," a cipher and, crucially, a cipher without *place*. How can "the thing itself" be "no thing" at all? As Recorde recalls in his verses on algebra,

One thyng is nothyng, the proverb is
Whiche in some cases doeth not misse.
Yet here by woorking with one thyng
Soche knowledge doeth from one roote spryng
That one thyng maie with right good skille,
Compare with all thyng.

That Shakespeare would have known Recorde's poem,[54] or any other works on what Dee called the use of "Aequation" to find out the value of unknowns is unlikely, but he knew well enough the proverb Recorde refers to, "One is no number." As noted in chapter 2, Shakespeare continually invoked the Renaissance idea that one, since it contains no plurality, is "no number," and thus "nothing"; it is just another denomination of zero. For example, Capulet asks Paris to compare his daughter's merit to others, hoping, against mathematical "reason," that "mine,

---

[53] Dollimore wisely comments that "in a world where pity is the prerequisite for compassionate action, where a king has to share the suffering of his subjects in order to 'care,' the majority will remain poor, naked, and wretched" (*Radical Tragedy*, 191).

[54] The verses are from Recorde, *Whetstone of Witte*, sig. B1v. It is doubtful that Shakespeare knew this work, which was never republished. The *Whetstone* is best known for having introduced the modern equal sign (=).

being one, / May stand in number, though in reck'ning none"
(1.2.32–33).[55] In the number world of Renaissance England, a man who
is a "cipher" has no value; he is one thing, or no thing, until he acquires
a "place" among men with whom he can measurably relate. "Unaccommodated man"—a poor, bare, forked animal—needs something more to
be counted among the "number" of society, as Lear and Gloucester recognize on the heath; the trouble is that he first must be counted among
the fellowship of man in order to be granted his addition. Although the
play advocates giving more to those with too little, it's not clear that Bedlam beggars have "enough" humanity to be counted as part of the fellowship of man to begin with; after all, men who cannot *feel* (for example, a pinprick) are not actually men, in *Lear*'s terms.[56] To number,
and to be numbered, is to be human, according to Recorde: "So that of
number, this may I justly say, it is the onely thing almost that separateth
man from beasts. He therefore that shall contemn number, declareth
himself as bruitish as a beast and unworthy to be counteth in the Fellowship of man."[57] The "Fellowship of man" may be one of the great revelations of *King Lear*, what it means to be counted human, according to
early seventeenth-century systems of human measurement. Recorde's
name for geometric distribution, tellingly enough, was the Rule of Fel-

---

[55] Karl Menninger, *Number Words and Number Symbols: A Cultural History of Numbers*,
trans. Paul Broneer (Cambridge, MA: MIT Press, 1969), has explained the history of
this mathematical idea:

> One, as the antithesis of Many, had already taken a special position as far back
> as the Table of Ordinal Concepts established by the Pythagoreans. Plato con-
> stantly emphasized this: Like the Now in time and the Point in space, the One
> among the numbers cannot be further subdivided. Hence it conceals within
> itself no plurality which it collects together into a unity, and since it is in this
> that the essence of number lies, One is not a number. Since, according to Eu-
> clid, "a number is an aggregate composed of units," One is itself not a num-
> ber, though it is the source and the origin (*fons et origo*) of all numbers.
> Throughout the Middle Ages no one thought differently. (19)

Menninger cites Michael Stevin, in 1585, as the first mathematician to assert that one
was a number after all. He proved it by saying that if we subtract a nonnumber from 3
then 3 remains, but since 3 − 1 = 2, one must be a number (20).

[56] *Lear* several times refers to the contemporary notion that madmen cannot "feel."
Edgar refers to the way they "[s]trike in their numbed and mortified arms / Pins,
wooden pricks, nails" (2.3.15–16), while Lear argues his sanity by noting, "I feel this
pin prick" (4.7.55).

[57] Recorde, "Preface to the Loyvyng Readers," in *Ground of Artes*. Compare Thomas
Hill's formulation, n. 45, above.

lowship.[58] According to the social mathematics of *King Lear*, "the rule of fellowship" dictates both what unites and what divides us, our "equalities"—in the Renaissance sense of the word.

For all his attention to the discourse of Renaissance mathematics and its applications to theories of socioeconomic justice, Shakespeare remains doubtful about the extent to which numbers can solve the problem of the relationship between what one is and what one gets, or should get, in proportion to it. In this, the playwright participates in a strain of Renaissance skepticism aimed specifically at human "calculations." Richard Pont, in his *New treatise of the right reckoning of yeares* (1599), uses "both an Arithmeticall & Geometricall proportion of numbers" to reckon the end of human history, since all "men wil reckon and counte by Arithmeticke all things." All men, however, will not count all things correctly: "[O]nelie the faithfull, (and they that put their confidence in God)" can "rightlie reckon."[59] Scheltco Geveren, in his treatise *Of the ende of this worlde* (1578), also grieved that neither true "Arithmetical proportion" nor "Geometric proportion" is "kept in this wicked world" but only in the world to come, "the eternall kyngdome of Christ, which with vpright iudgement, *and by equalities in euery respect,* he will establishe perpetually, and make it endure world without end."[60] In his *Six Bookes of the Commonweale* (1606), Jean Bodin offers a sustained critique of contemporary social mathematics, finding fault both with arithmetic and geometric systems of economic justice. The common people, Bodin observes, will always seek "Arithmetical Iustice," "the people still seeking after nothing more, than for equalitie in all things," while aristocrats will always favor "Geometricall proportion."[61] In fact, neither serves the economic interests of rich or poor:

> [W]e see the rich man by Geometricall proportion of Iustice, to be much more grieuously fined than the poor: and so contrariwise the Arithmeticall proportion of Iustice, in the imposing of

[58] Ibid., 168.

[59] Richard Pont, *New treatise of the right reckoning of yeares* (Edinburgh, 1599), 94.

[60] Scheltco Geveren, *Of the ende of this worlde, and seconde commynge of Christe* (London, 1578), 42, 44, emphasis added.

[61] Jean Bodin, *The Six Bookes of the Commonweale*, facsimile ed., ed. Kenneth Douglas McRae (Cambridge, MA: Harvard University Press, 1962), 755.

penalties and fines, to be the meanes for the rich to vndoe the
poore, and all vnder the colour of iustice.

As a further example, he observes,

> So the Surgeon which taketh of the rich man fiue hundred
> crownes to cut him of the stone, haply taketh of the poore porter
> no more but fiue: and yet for all that in effect taketh ten time
> more of the poore man than of the rich: For the rich man being
> worth fiftie thousand crownes, so payeth but the hundred part of
> his goods, whereas the poore man being but worth fiftie crownes,
> paieth fiue, the tenth part of his substance. Whereas if we should
> exactly keepe the Geometricall or Arithmeticall proportion
> alone, the patient should die of the stone, and the Surgion for
> lacke of worke starue.

Bodin proposes that economic fairness demands an admixture of the
two, a combined measure, which he terms "the true proportion Har-
monicall, and whereof no man hath yet spoken." He contrives a human
equation, a romantic "match," to illustrate his theory of social "harmon-
ics":

> [I]f a rich base woman marrie with a poore gentleman; or a poore
> gentlewoman with a rich common person: and she that in beauty
> and feature excelleth, vnto him which hath some one or other
> rare perfection of the mind: in which matches they better agree,
> than if they were in all respects equall.[62]

Bodin also attempts to clear up what he sees as a dangerous confusion in the rhetoric of human justice: the word "equal," he insists, is supposed to mean the "same," an arithmetical proportion of one-to-one,
and it is erroneous and misleading to deem "geometrical" equivalences
"equal" as well:

> The difference of the Geometricall and Arithmetical proportion,
> is in this to be noted, That in the proportion Arithmeticall are al-

---

[62] Ibid., 770, 783, 758.

wayes the selfe same reasons, and the differences equall: whereas in the Geometricall proportion they are alwayes semblable, but not the selfe same, neither yet equall: except a man would say, that things semblable are also equall; where were nothing else, but improperly to speake.

To speak thus "improperly" is a form of political sophistry and mystification, to proceed

as Solon did, who to gaine the hearts both of the nobilitie, and of the people of Athens, promised to make them lawes equall for all sorts of men: wherein the nobilitie and better sort of the people thought him to haue meant the Geometrical equalities, and the common people, the Arithmeticall; and so to haue been all equall: Which was the cause that both the one and the other by common consent made choyce of him for their lawmaker.

Although Bodin attempts to revolutionize the prevailing social arithmetics of his day by "adding" the current rules together and by resolving the rhetorical ambiguity surrounding the word "equality," he does not critique the ideological turn to numbers itself. In our determinations of economic fairness, he maintains, "[I]t is [still] needfull for vs to borrow the principles of the Mathematicians."[63]

Rather than tweaking the numbers in an effort to get it "right," Shakespeare remains uncertain as to whether mathematical measures, of any kind, are apt to the purpose of determining justice. At the end of *King Lear*, Cordelia suggests the futility of efforts to "match" human virtue, to "measure" human deserving:

O thou good Kent, how shall I live and work
To match thy goodness? My life will be too short,
And every measure fail me.
(4.7.1–3)

Shakespeare no doubt shared Cordelia's sense that people should "get" what they deserve, yet, arguably, the world of *Lear* is one in which "every

---

[63] Ibid., 758, 757, 770, 755.

(mathematical) measure" that's applied "fails" in this very regard. Perhaps the common, modern critical response to *Lear*'s conclusion—that matters between rich and poor still do not stand fully "corrected"—reflects a degree of equivocation on Shakespeare's part, an uneasiness regarding the human "remainders" of the play's tragic divisions, those factored out as "undeserving" poor. Like Bodin, Shakespeare had misgivings about particular applications of mathematics to human problems, as they function in the play. With Seneca, Shakespeare may have questioned the value of human "numbering" altogether:

> The mathematician teaches me how to lay out the dimensions of my estates: but I should rather be taught how to lay out what is enough for a man to own. He teaches me to count, and adapts my fingers to avarice; but I should prefer him to teach me that there is no point in such calculations, and that one is none the happier for tiring out the bookkeepers with his possessions—or rather, how useless property is to any man who would find it the greatest misfortune if he should be required to reckon out, by his own wits, the amount of his holdings. What good is there for me in knowing how to parcel out a piece of land, if I know not how to share it with my brother? What good is there in working out to a nicety the dimensions of an acre, and in detecting the error if a piece has so much as escaped my measuring-rod, if I am embittered when an ill-tempered neighbor merely scrapes off a bit of my land? The mathematician teaches me how I may lose none of my boundaries; I, however, seek to learn how to lose them all with a light heart.[64]

Certainly, the mathematical "measuring rods" of *King Lear*, from the division of the kingdom to its final dispensations, teach nothing if not how to lose everything, to feel only the incalculable "*weight* of th[e] sad time" of Shakespeare's tragedy (5.3.324, emphasis added).

The Renaissance idea that there are two equalities—one arithmetic, one geometric; one based on how we are the same, and the other on how we

---

[64] Lucius Annaeus Seneca, *Epistles 66–93*, trans. R. M. Gummere (Cambridge: Harvard University Press, 1992), vol. 5, epistle 88.

are different—may sound to us suspiciously like the Orwellian axiom that "all animals are equal, but some are more equal than others." There's no apologizing for the ideological inconsistencies regarding ideas about economic distribution in *King Lear*, or for the errors in the mathematical proofs offered in support of them. It is unlikely, though, that it all added up wrong in the seventeenth century. What looks to us now as faulty logic, if not bad faith, is all too common in Renaissance discourses on human "relations."

In his essay "Of the Inequality That Is between Us," Montaigne, for example, maintains that "there is more distance from a given man to a given man than from a given man to a given animal," citing Terence, "How far one man excels another!"[65] Yet Montaigne further argues that

> [i]f we consider a peasant and a king, a nobleman and a plebeian, a magistrate and a private citizen, a rich man and a pauper, there immediately appears to our eyes an extreme disparity between them, though they are different, so to speak, only in their breeches. (191)

What establishes our common human nature, for Montaigne, are the sensations of the body and mind; of the monarch, he asks,

> Do fever, migraine, and gout spare him any more than us? When old age weighs on his shoulders, will the archers of his guard relieve him of it? When the terror of death paralyzes him, will he be reassured by the presence of the gentlemen of his chamber? When he is in a jealous and capricious mood, will our taking off our hat set him right? . . . Where body and mind are in bad shape, what is the use of these external advantages, seeing that the slightest pinprick, or passion of the soul, is sufficient to deprive us of the pleasure of being monarch of the world? (191–92)

Montaigne writes further of the problem of a king's having an abundance, too much. But "above all," he says, the monarch

---

[65] Michel de Montaigne, "Of the Inequality That Is between Us," in *The Complete Essays of Montaigne*, trans. Donald M. Frame (Stanford: Stanford University Press, 1957), 189–98, quote at 189. All further references to this essay come from this edition and are cited parenthetically in the text.

finds himself deprived of all mutual friendship and society, wherein consists the sweetest and most perfect fruit of human life. For what testimony of affection and good will can I extract from a man who, willy-nilly, owes me everything he can do. . . . My elevation has placed me outside of human association: there is too much disparity and disproportion. (195)

How is it possible to reconcile Montaigne's idea that we are essentially unequal with his following insistence that we are, also in some essential way, all equal and that there is therefore too much "disparity and disproportion" between us? Is his essay an indictment of the innumerable social differences that divide us? I don't think so, any more than I think that *Lear* is. But isn't the "human association" that the monarch lacks to be founded on mutuality, the leveling of an artificial "elevation"? The answer here is yes and no, as it must be in *King Lear*. Although the polity is to be based on the king's understanding of his physical and spiritual proximity to all men, he can only be king by virtue of superior merit.[66] Martin Luther had also suggested as much, along with the mathematical concept of "place" that accredits it:

> To the counting master all counters are equal, and their worth depends on where he places them. Just so are men equal before God, but they are unequal according to the station in which God has placed them.[67]

Montaigne concludes his essay by citing Anacharsis: "The best government would be one in which, all other things being equal, precedence was measured out according to virtue, and rejection according to vice" (196). All other things being equal, there will be precedence—again, there is no contradiction here for Montaigne, who, after all, treats human equality in an essay on human *inequality*, without remark, almost as if to him they were the same thing. And for many Renaissance writers, drilled in the mathematical rhetoric of "equality," they were.

---

[66] Compare Gonzalo in *The Tempest*, who imagines a commonwealth based on a leveling of political and economic differences and yet, as Sebastian points out, "he would be king on't" (2.1.157).

[67] Quoted in Alfred W. Crosby, *The Measure of Reality: Quantification and Western Society, 1250–1600* (Cambridge: Cambridge University Press, 1997), 45.

The ways that equalities are weighed in *King Lear*, however specious they may seem to us today, nevertheless reveal something of the history of the rhetoric of "equality" in the West. That rhetoric is always in danger of equivocation, simply because theories of identity—themselves inconstant, contingent, and culture specific—lie, implicitly, at the core of theories of human equality. Standards of social comparison are not fixed and self-evident but rather variable, depending on what is placed on either side of the equation.[68] As social philosophers have reminded us,

> Our language permits us to view a man in several ways, as his moral self, or as his moral self combined with his intelligence, or as his moral self, intelligence, and his social position, including wealth. The way in which we conceive the self amounts to formulating criteria of relevance for differential treatment, and one's system of goods distribution will vary greatly, depending on the criteria chosen (and their order of priority).[69]

Some of the confusion regarding the question of social and economic differences in *King Lear* is inherent in the notion of equality itself, as it has been applied historically to people. Even the idea of "equality by nature" is not a self-evident fact but a value; it is, after all, easily challenged by the fact of physical differences among us, as even Edmund is aware.[70]

---

[68] S. I. Benn and R. S. Peters, "Justice and Equality," in *The Concept of Equality*, ed. William T. Blackstone (Minneapolis: Burgess, 1969), have argued the radical contingency of the value "human equality":

> Out of context, "equality" is an empty framework for a social ideal; it has content only when particularized. That is why it is a mistake to think of history as a movement towards ever greater equality, in which one distinction after another is being torn down; as though the word stood for some essential idea which assumed a different shape in each generation, but which remained in some fundamental way the same, or of which every later manifestation somehow embraced and transcended all the earlier ones. For as fast as we eliminate distinctions, we create new ones—the difference being that the ones we discard we consider unjustifiable, while the ones we create seem reasonable. (62)

[69] William T. Blackstone, "Introduction," in Blackstone, *Concept of Equality*, ix.

[70] Despite his claim to be "equal" to Edgar in nature, Edmund immediately follows by suggesting that the lust that engendered him gives him greater strength, or "more composition" (2.1.12), than those born under the law. For a fuller discussion of Renaissance notions of a human "equality by nature" and why even nature does not provide a stable criterion for assessing human relations, see chapter 3.

Human equality is not, by any measure, an empirical fact, since there's no empirical data, psychological, biological, or sociological, that can prove (or disprove) it.

Why, then, is the sense of "equality" as a single, essential, transcendent idea about humanity so deeply ingrained in the Western imagination? Perhaps it's because we continue to conceive of the notion of equality as a mathematical one. Because mathematical equality is adducible and incontrovertible, it seems to us that human equality is or should be as well. With human figures, however, as opposed to arithmetical ones, equality will always depend on which of the particular qualities of men and women, and the lives they lead, are to be counted and compared. The inevitable selectivity of that process, moreover, creates justice systems that must acknowledge if not celebrate, with Goneril, "the difference between man and man." If we share, with Shakespeare's *King Lear*, the idea of our being "created equal," we also share a habit of "creating equals," rhetorically, to suit the dreams we have for ourselves and our societies.[71]

[71] For a brilliant treatment of our modern rhetorics of equality, see Peter Westen, *Speaking of Equality: An Analysis of the Rhetorical Force of "Equality" in Moral and Legal Discourse* (Princeton: Princeton University Press, 1990). Westen considers, for example, how "the concept of equality does not itself contain criteria for judging standards of comparison; it presupposes them" (121).

THE LESBIAN RULE OF
MEASURE FOR MEASURE

A MID-SIXTEENTH-CENTURY EMBLEM by Petrus Costalius figures the *bonus iudex*—the good judge—as one who determines what is right *in normam lesbiam*, "by the Lesbian rule."[1] As the emblem depicts it, the *norma Lesbia* or *regula Lesbia* was an ancient carpenter's measure flexible enough to adapt to the surface of irregular stones. Aristotle, in his *Ethics*, may have been the first to imagine the aptness of the Lesbian rule as a metaphor for the application of *epieikeia*, or "equity," to certain cases in law:

> [L]aw is always a general statement, yet there are cases which it is not possible to cover in a general statement. . . . This is the essential nature of the equitable: it is a rectification of law where law is defective because of its generality. In fact this is the reason why things are not all determined by law: it is because there are some cases for which it is impossible to lay down a law, so that a special ordinance becomes necessary. For what is itself indefinite can only be measured by an indefinite standard, like the leaden rule used by Lesbian builders; just as that rule is not rigid but can be bent to the shape of the stone, so a special ordinance is made to fit the circumstances of the case.[2]

---

[1] Petrus Costalius [Pierre Coustau], *Pegma* (Lugdini: Apud Matthiam Bonhomme, 1555).

[2] Aristotle, *Nichomachean Ethics*, trans. H. Rackham, Loeb Classical Library (London: William Heinemann, 1947), V.10.3.

When laws fail to "fit" the shape of a case, equity—a "special kind of Jus-tice"—re-forms them; by bending the law, paradoxically, equity "recti-fies" it. For Aristotle as for many of his Renaissance heirs, equity is not an alternative measure with its own specific shape and dimensions but rather a universal "measure of all things." It is the use of an indefinite standard or "irregular rule" that distinguishes the good judge—even, for Aristotle, the superior judge—for "Justice and equity are therefore the same thing, and both are good, though equity is the better."[3] In this chapter, I will examine the rhetoric of measurement in Renaissance legal discourse—from straight "rules" to "lines" of authority, from the "level" of law to the "scales" of justice—focusing on the claims of com-peting measures to gauge or guarantee equable human relations.[4] Shakespeare's investigation of the metaphor of "measurement," as ap-plied to judgment, exposes the potential for "mismeasure" or "misrule" in our assessments of what is "right" (from Latin *rectus*, "straight") in human behavior.[5]

The notion of making judgments "by the Lesbian rule" was proverbial by the sixteenth century. Although most often invoked as a method for making legal decisions, it is also used to describe an alternative to "lit-eral" interpretation more generally. Reading by the Lesbian rule, like any "equitable" judgment, meant deferring the words of the work, sus-pending the "letter" of the text. Thus when Thomas Sackville weighed in on the sixteenth-century debate over translating Cicero, he compared

---

[3] Ibid., V.10.2.

[4] See chapter 4 for a discussion of the Renaissance rhetoric of equality as it was ap-plied to matters of economic "equality."

[5] Tempting as it might be to imagine that the word *Lesbian*, in these contexts, refers to female homosexuality, we must recognize that Renaissance writers did not use the term in this sense. I have tried to avoid interpretations of the "Lesbian rule" that are based in verbal anachronism. Recent work in queer philology, however, has shown how Renaissance uses of the word *straight* intersect in important ways with the history of sexual behavior and sexual identity. As Laurie Shannon has argued, Renaissance "homonormativity"—the principle that like naturally tends toward like—implicitly identified what is "straight" with same-sex attraction. See Shannon, "Nature's Bias: Renaissance Homonormativity and Elizabethan Comic Likeness," *Modern Philology* 98.2 (November 2000): 183–210; and idem, *Sovereign Amity: Figures of Friendship in Shakespearean Contexts* (Chicago: University of Chicago Press, 2002). The Lesbian rule does not refer in any direct way to Renaissance (homo)sexuality, but insofar as it rep-resents a rule that conforms to what it measures, that bends to the same shape, it of-fers an intriguing example of Renaissance "homonormativity," or "norms" based on "sameness."

In normam Lesbiam.
Bonus iudex.

Aspice vt in partes semper tornatilis omnes,
Se subiectam operi Lesbia norma facit.
Intus sæpe latet scriptæ sententia legis,
Quæ solet in tota iudicis esse manu.
Hanc ille in varijs rerum variare figuris
Debet, non vnum semper, idémque sequi.

E iiij

those who pored over the letter of the ancient rhetorician's texts with those who attempted to do justice to his meanings:

> They do badly and very unjustly who always weigh Cicero by the scale of Nizzoli rather than by the lesbian rule, since the imitation of Cicero consists not so much in the hunting of words, as in a certain proper weight and number of meanings.[6]

The inscription for Costalius's emblem, *in normam Lesbiam,* identifies the good judge as a generous *reader* of law:

> Look at how the turned Lesbian rule,
> Conforms itself in all directions to the work.
> Often the meaning of a written law hides within—
> Which is usually entirely in the hand of the judge.
> He must vary [the rule] according to the shape of shifting things,
> Not always follow one and the same way.

> ([A]spice vt in partes semper tornatilis omnes,
> Se subiectam operi Lesbia norma facit.
> Intus saepe latet scriptae sententia legis,
> Quae solet in tota iudicis esse manu.
> Hanc ille in varijs rerum variare figuris
> Debet, non vnum semper, idémque sequi.)

Renaissance writers often reinterpreted the classical notion of the superiority of equity to an unwavering adherence to the law as a type of the Pauline celebration of the "spirit" of the text as against its "letter." In its association with Christian mercy, equity, as a mode of reading, promised access to the true "life" of the work. Aristotle's promotion of equity over law thus came to be identified with a Christian *progress* from literal to spiritual understandings.

Despite the legacy of Aristotle, however, and even more, a Christian insistence on equity as superior to law in determinations of "truth," adju-

---

[6] Thomas Sackville, "Amplissimo," preface to Castiglione's *The Courtier* (London, 1577). Mario Nizzoli was a chief exponent of the sixteenth-century view that Cicero's *language* should be imitated above all else. See Izora Scott, *Controversies over the Imitation of Cicero in the Renaissance* (Davis, CA: Hermagoras Press, 1991).

dicating "by the Lesbian rule" was not always deemed a legitimate proceeding. Erasmus included the phrase "By the lesbian rule" in his *Adagia*,
with the following indictment:

> This is said when things are done the wrong way round, when the-
> ory is accommodated to fact and not fact to theory, when law is
> suited to conduct, not conduct corrected by law; or when the ruler
> adapts himself to the behaviour of the populace, though it would be
> more fitting for common people to conduct their lives according to
> the will of the prince; if only the prince himself has right conduct
> before his eyes as his rule and aim. Aristotle mentions this adage in
> his *Ethics*, book 5: "For the rule of what is indefinite is also indefi-
> nite, like the leaden rule used in Lesbian architecture; the rule
> changes to fit the shape of the stone and does not remain a rule."[7]

Elsewhere, Erasmus invoked the Lesbian rule in reference to any willful,
erroneous interpretation, attributing its use, for example, to those who
would "rather distort Scripture than correct human judgments by the
rule of Scripture." For Erasmus, the Lesbian rule perverts rather than
promotes the discovery of truth by confusing the criterion of measure
with what is measured; true judgment depends on a consistent, invariable, and authorized rule. Thus alongside the adage "By the lesbian
rule," he cites its happier counterpart, "By rule":

> In the best authors we often find *Ad amussim*, By rule (with, *exam-
> ussatim, amussatim*) put for "with the utmost diligence," "with the
> most precise care." Gellius, book 1, chapter 4: "He inspected all
> these ancient writings with such care, weighing their merits and
> investigating their short-comings, that you would say his judgment
> was made exactly by rule." . . . This is taken from stone-cutters or
> carpenters, who test the evenness of their work with the familiar
> measuring-line.[8]

In the Renaissance, *both* the "straight" rule and the Lesbian rule, as measures, claimed "evenness" or equality as the result of their operations.
The Renaissance debate over which measure was more "just" depended,

[7] Desiderius Erasmus, *Adagia*, in *The Collected Works of Erasmus* (Toronto: University of
Toronto Press, 1982), vol. 31, I.v.93.
[8] Ibid., I.v.90.

as we shall see, on divergent interpretations of the principle we now call "equality under the law."

Although the *Oxford English Dictionary* cites the first usage of the phrase "legal measure" as occurring in the eighteenth century, the identification of the law with instruments of measure is an ancient one. By the sixteenth century, laws were commonly referred to as "rules" (the early modern English word for what we now call "rulers," as in Cleopatra's scorn for "[m]echanic slaves / With greasy aprons, *rules*, and hammers" [*Antony and Cleopatra*, 5.2.209–10, emphasis added]) or "lines" (an early modern variant of "measuring lines" or, again, our "rulers"). Thus Angelo in *Measure for Measure* determines to rule Vienna "with full *line* of his authority" (1.4.56, emphasis added). When Sir Walter Blunt, in the *First Part of Henry IV*, warns Hotspur he will oppose him "[s]o long as out of limit and true rule / You stand against annointed majesty," he invokes the common notion of a rule as a measure of extent (4.3.39–40). Other linear measures frequently used in reference to the law include the level ("an instrument which indicates a line parallel of the horizon") and the square ("an instrument for determining, measuring, or setting out right angles, or for testing the exactness of an artisan's work").[9] Phillip Stubbes, for example, defined human justice in such terms: "The lawe in itself, is the square, the levell, and rule of equitie and justice."[10] When Trinculo jests that he and Stephano "steal by line and level" (*Tempest*, 4.1.239), the joke hinges on the paradox that they are breaking the law, as it were, "by the rules." In reference to his own, personal "misrule," Antony insists, "I have not kept my square, but that to come / Shall all be done by the rule" (2.3.6–7). The coincidence of Latin *regula* as "rule" meaning ruler or measuring rod and "rule" meaning principle of order or conduct may not have been the ultimate source of the notion of legal "measures," as Aristotle's example makes clear, but it no doubt helped fix the idea in the Renaissance imagination. Timon of Athen's demand for "regular justice" (5.4.61), for example, speaks directly to Shakespeare's engagement with the "rule" of law.

Shakespeare and his contemporaries also inherited an identification of justice with scales or balances, instruments that determine "even" or "equal" weight. The *Oxford English Dictionary* cites Shakespeare himself as the first to use the word "scale" as an attribute of justice, but the meta-

---

[9] *Oxford English Dictionary*, s.v. "level" and "square."
[10] Ibid., s.v. "level."

phor, once again, has its sources in antiquity. Although references to Astraea, the Roman goddess of Justice, are relatively rare in classical texts, she came to emblematize the ideal of justice in early modern literature, especially the relationship between divine and human justice. Natalis Comes's *Mythologiae* (1581) may provide the most extended early modern exposition of the story. According to the *Mythologiae*, during the golden age, perfect justice held sway and laws were "written" in men's hearts. As the golden generations of men gave way to inferior ones, Astraea fled the earth, leaving a book of laws behind her to remind men of the way to govern themselves. Astraea, who holds the "balance" of Justice, is metamorphosed as the constellation Virgo, and her scales the neighboring constellation Libra. The "scales" of Justice are a commonplace of Renaissance writing. Erasmus construes the adage "Stateram ne transgrediaris" to mean "Exceed not the balance, that is, You shall not do anything which is not just and right. For the balance was commonly held in old days to be a symbol of equity, as is shown by that Doric proverb: As fair as a pair of scales."[11] Edmund Spenser's extended examination of the scales of justice in book 5 of the *Faerie Queene* is representative in its examination of law and its measures, and its polemic of "equality" is explored later in this chapter. In the *Faerie Queene* as in many other Renaissance texts, the classical notion of "weighing" as a procedure for justice was often incorporated into Christian accounts of judgment, the final "reckoning" of sin. Giles Fletcher, for example, imagines "Christs Victorie in Heaven" as a debate between the Virgin goddess, who "in one hand a paire of even scoals [*sic*] . . . weares," and Mercy; in his verses, the classical terms of judgment match or outweigh the biblical ones.[12] Thomas Tymme, who foretold how "[t]he deeds of men . . . are . . . to be examined by Gods level and line,"[13] describes Christ's sacrifice as a "weighty" act that balances the sins of mankind: "Our sinnes being weighed in the ballaunce of the iustice of GOD, were found to bee so weightie, and of so great importance, that his wrath coulde neuer haue beene appeased towardes us, but by the death of his onely Sonne, which maketh full satisfication."[14]

Shakespeare makes numerous references to scales and balances in his poems and plays, sometimes as "real" instruments of measure but more

11 Erasmus, *Adagia*, I.i.2.
12 Giles Fletcher, *Christs Victorie in Heaven* (Cambridge, 1610), 10.8.
13 *Oxford English Dictionary*, s.v. "level."
14 Thomas Tymme, *The Court of Conscience* (London, 1605), 247.

often as metaphors for justice or human judgment more generally. The scales on hand to weigh Antonio's flesh in the *Merchant of Venice* serve both purposes: they are a literal reminder that Portia's judgment will depend on how she handles this instrument of correction and commensuration (see chapter 3). King Henry VI assures Gloucester that he will "poise [Gloucester's] cause in justice' equal scales / Whose beam stands sure, whose rightful cause prevails" (*2 Henry VI*, 2.1.200–201). The bumbling constable Dogberry urges his charges, "Come, sir, if justice cannot tame you, she shall ne'er weigh more reasons in her balance" (5.1.206–7). Prince Hal thanks the Chief Justice for his counsel: "You are right justice, and you weigh this well; / Therefore still bear the balance and the sword" (*2 Henry IV*, 5.2.102–3). Laertes vows to avenge Ophelia's madness by means of a proportionate reprisal: "By heaven, thy madness shall be paid with weight / [Till] our scale turn the beam" (*Hamlet*, 4.5.157–58).

Shakespeare, however, understands above all how easily the scales of judgment may be tipped, how "equality" may be contrived rather than indifferently ascertained. As the Porter in *Macbeth* suggests, for example, scales may be manipulated by "equivocation": "Faith, here's an equivocator, that could swear in both the scales against either scale" (2.3.8–9). If, like the Porter, Shakespeare believed that no man can "yet . . . equivocate to heaven" (2.3.11), he was not so sure about the case on earth. For Shakespeare, human justice deals, all too often, in "equivocations" (from Latin *aequivocare*, "to call by the same name") rather than evenhandedness, rhetorical rather than "real" equalities. Shakespeare's skepticism toward legal measures centers on doubts about the criteria that authorize them, the problem of verifying the "legitimacy" of the law, as a vehicle of human judgment.

Plato's *Laws* begins with a question about origins:

ATHENIAN: To whom do you ascribe the authorship of your legal
    arrangements, Strangers? To a god or to some man?
CLINIAS: To a god, Stranger, most rightfully to a god.[15]

---

[15] Plato, *Laws*, trans. R. G. Bury, Loeb Classical Library (Cambridge, MA: Harvard University Press, 1926), B 624.

From the start, philosophers of law foreground the sources of human legal systems, especially the basis for their authority. How is it possible to certify the "rightness" of laws themselves? On what grounds can one guarantee the "justice" of justice?[16]

The earliest legal philosophers make one thing clear: if all laws derived from the gods, there would be no need to question the standards for legal authority. The problem—implicit in the Athenian's opening question—is what happens when the source of law, instead, is "some man," that is, when "man is the measure" of justice. From Plato forward, human law is continually justified by presuming its basis in, or authorization by, divine law. There are two prescribed ways of receiving divine law—either directly, in the form of written commandments of the gods, or indirectly, through the intercession of "natural law," that is, the incarnation of divine law as human nature. For classical writers, "natural law" is an undefined but innate and universal code of human conduct, based in human reason; for Christians and Christian humanists, it derives both from reason and conscience. Either way, "natural Law [is] essential justice, justice itself, the origin and test of all positive laws, and the ultimate measure of right and wrong."[17] In the case of divine injunction or scripture, measuring the validity of human law against the gods' is a relatively simple, unequivocal procedure. We are given no reason to doubt the justice of Clarence's appeal to his murderers that they countermand the warrant for his death:

> the great King of kings
> Hath in the table of his law commanded
> That thou shalt do no murther. Will thou then
> Spurn at his edict, and fulfill a man's?
> (*Richard III*, 1.4.195–98)

Although natural law too is often invoked as a clear, inviolable criterion for the evaluation of human law, it is never that straightforward: the stipulation of a universal, "standard" human nature is continually challenged by the particularity and partiality of human will. Shakespeare, es-

---

[16] R. S. White, *Natural Law in English Renaissance Literature* (Cambridge: Cambridge University Press, 1996), 1.
[17] Ibid.

pecially, would return again and again to the laws that men impose on others. Thus Valentine describes his raucous company as "uncivil" men "that make their wills their law" (*Two Gentlemen of Verona*, 5.4.17, 14), and Alcibiades censures the Senators:

> Till now you have gone on and fill'd the time
> With all licentious measure, making your wills
> The scope of justice.
> (*Timon of Athens*, 5.4.3–5)

The variability of human nature(s) leaves open the possibility that the scope of its justice may prove "licentious"—out of all legal measure.

The ideal of legal universality and invariability also left philosophers, from the start, with an apparent paradox: if human laws are grounded in divine ones, what accounted for the historical and cultural variations among them? Moreover, on what basis are laws subject to change? Alongside ancient claims for the "natural" grounds of law there were in fact always concessions to its conventionality, often made in the service of arguments for legal reform. Aristotle explains that

> Political Justice is of two kinds, one natural, the other conventional. A rule of justice is natural that has the same validity everywhere, and does not depend on our accepting it or not. Some people think that all rules of justice are merely conventional, because whereas a law of nature is immutable and has the same validity everywhere, as fire burns both here and in Persia, rules of justice are seen to vary. That rules of justice vary is not absolutely true, but only with qualifications. Among the gods indeed it is perhaps not true at all; but in our world, although there is such a thing as Natural Justice, all rules of justice are variable. But nevertheless there is such a thing as Natural Justice as well as justice not ordained by nature.[18]

Fortunately, according to Aristotle, the distinction between natural and conventional law is unmistakable: "[I]t is easy to see which rules of justice, though not absolute, are natural, and which are not natural but

---

[18] Aristotle, *Nichomachean Ethics*, V.vii.1–3.

legal and conventional."[19] Yet the "truth" of natural law, clearly enough, was not always self-evident. As it happened, for all his own emphasis on legal certainty, Aristotle is often cited by Renaissance writers as a source for the idea of the *fallibility* of human law. Abraham Fraunce, in his *Lawiers Logike* (1588), was among many who reminded his readers of "what Aristotle sayth of the imperfection of all Lawes."[20]

Cicero proposed a way of resolving the contradiction of "natural" laws that proved invalid. In *De legibus*, Cicero insists on reason as the source of law:

> Law is the highest reason, implanted in Nature, which commands what ought to be done and forbids the opposite. This reason, when firmly fixed and fully developed in the human mind, is Law. . . . [T]hen the origin of Justice is to be found in Law, for Law is a natural force; it is the mind and reason of the intelligent man, the standard by which Justice and Injustice are measured [ea iuris atque iniuriae regula].[21]

Although man's reason is the measure, God is the ultimate standard of justice; his law authorizes human "rule":

> True law is right reason in agreement with nature; it is of universal application, unchanging and everlasting. . . . It is a sin to try to alter this law, nor is it allowable to attempt to repeal any part of it, and it is impossible to abolish it entirely. . . . And there will not be different laws at Rome and at Athens, or different laws now and in the future, but one eternal and unchangeable law will be valid for all nations and all times, and there will be one master and ruler, that is, God, over us all, for he is the author of this law, its promulgator, and its enforcing judge.

Cicero attempts to resolve the paradox of legal reform, that is, the contradiction in asserting human laws to be based in eternal, divine law, yet somehow still subject to change, by making a critical ontological distinc-

[19] Ibid., V.vii.4.

[20] Abraham Fraunce, *The Lawiers Logike* (London, 1588), sig. Aa2v.

[21] Cicero, *De legibus*, in *De re publica; De legibus*, trans. Clinton Walker Keyes, Loeb Classical Library (London: William Heinemann, 1928), I.vi.18.

tion: "bad" laws, laws that must be altered or suspended, were never laws
to begin with:

> [T]hose who formulated wicked and unjust statutes for nations,
> thereby breaking their promises and agreements, put into effect
> anything but "laws." . . . [I]n the very definition of the term "law"
> there inheres the idea and principle of choosing what is just and
> true. . . . Therefore Law is the distinction between things just and
> unjust, made in agreement with that primal and most ancient of
> all things, Nature . . . and in conformity to Nature's standard.[22]

"True" law can be measured by the standard of Nature; indeed, no other
standard may logically apply:

> [I]f [instead] . . . everything is to be tested by the standard of util-
> ity [utilitate omnia metienda sunt] then anyone who thinks it will
> be profitable to him will, if he is able, disregard and violate the
> laws. It follows that Justice does not exist at all, if it does not exist
> in Nature.[23]

Augustine would later cite Aristotle, perhaps via Cicero ("an unjust law is
not a law at all"), as would Thomas Aquinas (citing Aristotle, "A tyranni-
cal law, because not according to reason, is not strictly speaking a law,
but rather a kind of perversion of law").[24] By refusing some laws the status of law altogether, these theorists found a way to preserve the oldest
criterion for judging "justice"—its unchanging truth.

The influence of Aquinas's philosophy of law in the *Summa theologiae*
is especially significant here, since Aquinas emphasizes the concept of
law as that which both measures and is measured by truth. In defining
the "essence of law," Aquinas explains that "[l]aw is a kind of rule and
measure of acts. . . . Now the rule and measure of human acts is reason,
which is the guiding principle of human acts." Laws are to be understood

---

[22] Cicero, *De re publica*, in *De re publica; De legibus*, III.xxii.33; II.v.11, 12–13.
[23] Cicero, *De legibus*, I.xv.42.
[24] Augustine citing Aristotle, as quoted in Mark Tebbit, *Philosophy of Law: An Introduction* (London: Routledge, 2000); Thomas Aquinas, *Summa theologiae*, in *St. Thomas
Aquinas: Political Writings*, ed. and trans. R. W. Dyson (Cambridge: Cambridge University Press, 2002), IaIIae 92, p. 98.

as measures that apply to acts but also to the "rules" that organize and animate rational systems:

> Since law is a kind of rule and measure, it is said to be "in" something in two ways. In one way, as in that which measures and rules; and because ruling and measuring are proper to reason, it follows that, in this way, law is in reason alone. In another way, as in that which is measured and ruled; and in this way law is in all those things which are inclined to something by reason of some law.

This applies to people as well as to things: "[L]aw is present in someone not only as in one who rules, but also, by participation, as in one who is ruled. And it is in this latter way that each man is a law unto himself." Man is a "law unto himself," in other words, insofar as he is "ruled" by above, that is, by divine measures. Our own laws are not always certain measures:

> [T]hanks to the uncertainty of human judgment, especially in contingent and particular matters, it happens that different people judge human acts in different ways; and from this fact different and contrary laws arise. . . . And so human laws cannot have that infallibility which the demonstrated conclusions of the sciences have; nor is it necessary for every measure to be entirely infallible and exact, but only to such a degree as is possible.

The certainty of human judgment, however, can be verified by reference to a higher standard: "In order . . . that man might know without any doubt what to do and what to avoid, it [is] necessary for him to be directed in his proper acts by a law divinely given; for it is clear that such a law cannot err." Here he cites Augustine, "We see a law above our minds, which is called truth."[25]

For Aquinas, God is the final measure of man's measures:

> Law, because it is a rule and measure, can be in something in two ways, as stated above. In one way, as in that which rules and measures; in another way, as in that which is ruled and measured, for

---

[25] Aquinas, *Summa theologiae*, IaIIae 90, pp. 77, 70, 81; 91, pp. 90, 101.

something is ruled or measured in so far as it participates in rule or measure. Hence, since all things subject to Divine providence are rules and measured by the eternal law . . . it is clear that all things participate to some degree in the eternal law.

It would be hard to exaggerate how often Aquinas returns to this premise, as a way to proportion man's law to God's:

[T]he form of everything that is ruled and measured must be consistent with what rules and measures it. Now human law must answer to both these conditions, since it is both something directed to an end, and it is a kind of rule or measure ruled or measured in turn by a higher measure. And this higher measure is twofold, namely, the Divine law and the law of nature.[26]

Yet despite his "twofold" warranty, Aquinas backpedals, just a bit, with regards to the kinds of truths that legal measures have access to:

Human reason in itself is not the rule of things; but the principles which nature has implanted in it are general rules and measures of all things relating to human activity. It is of these things, and not of the nature of things in themselves, that natural reason is the rule and measure.

It turns out that there is not one but two kinds of "truths"—the nature of things in themselves, and the nature of "things relating to human activity." He cites Aristotle's notion that the mind is "measured" by what it apprehends:

The reason of the Divine intellect does not stand in the same relation to things as the reason of the human intellect does. For the human intellect is measured by things: that is, a human concept is not true simply in itself, but is said to be true by reason of its correspondence with things, for an opinion is true or false according to what a thing is or is not. But the Divine intellect is the measure

---

[26] Ibid., IaIIae 91, p. 86; 94, p. 132.

of things, since each thing has truth in it in so far as it represents the Divine intellect.[27]

There is nothing intrinsically true about human law; as a measure measured by what it measures, it operates within a confined, closed circle of human apprehension, cut off from the compass of divine truth.

Of Renaissance legal philosophers, Richard Hooker undertakes the most important reevaluation of Aquinas's notions of legal measurement. His discourse on the nature of law begins with Aquinas's formulation:

> Human laws are measures in respect of men whose actions they must direct; howbeit such measures they are, as have also their higher rules to be measured by, which rules are two, the law of God, and the law of nature.

Following Aquinas further, Hooker identifies reason as divine law incarnate in man: "Wherefore the natural measure whereby to judge our doings, is the sentence of Reason, determining and setting down what is good to be done."[28] However, although "the mind of man desireth evermore to know the truth according to the most infallible certainty which the nature of things can yield," reason does not always provide the "straightest" measure of truth:

> Where is then the obliquity of the mind of man? His mind is perverse, kam [*sic*], and crooked, not when it bendeth itself unto any of these things, but when it bendeth so, that it swerveth either to the right hand and to the left, by excess or defect, from that exact rule whereby human actions are measured. The rule to measure and judge them by, is the law of God.

Hooker sets up the old distinction between "law[s] either natural and immutable, or else subject unto change, otherwise called positive law" as the basis for his elaborate program for a new, "ecclesiastical polity" with reformed laws and systems of rule.[29]

---

[27] Ibid., IaIIae 91, p. 88; 93, p. 103.
[28] Richard Hooker, *Of the Laws of Ecclesiastical Polity*, in *The Works of Mr. Richard Hooker*, 3 vols., ed. Rev. John Keble (Oxford: Clarendon Press, 1888), 1:381, 232.
[29] Richard Hooker, "A Learned Sermon of the Nature of Pride," in *Works*, 3:599–600.

Although Hooker insists on grounding positive law in natural law, just as his predecessors had done, he explains that the relationship between them has long been misunderstood:

> The difference between which two undiscerned hath not a little obscured justice. It is no small perplexity which this one thing hath bred in the minds of many, who, beholding the laws which God himself hath given, abrogated and disannulled by human authority, imagine that justice is hereby conculcated; that men take upon them to be wiser than God himself. . . . The root of which error is a misconceit that all laws are positive which men establish.[30]

For Hooker, the important distinction to make, as far as the question of legal reform is concerned, is not between God's laws, which are natural, and man's laws, which are conventional, but between what is "natural" and what is "positive" among the measures of men. Crucially, the imperfections of the law need not lead inexorably to skepticism about man's ability to make the necessary corrections:

> We cannot therefore be persuaded that the will of God is, we should so far reject the authority of men as to reckon it nothing. . . . [M]en suffer themselves in two respects to be deluded; one is, that the wisdom of man being debased either in comparison with that of God, or in regard of some special thing exceeding the reach and compass thereof, it seemeth to them . . . as if simply it were condemned.[31]

It's illogical to indict human judgment altogether for the sake of bad laws; after all,

> I would gladly understand how it cometh to pass, that they which so peremptorily do maintain that human authority is nothing worth are in the cause which they favour so careful to have the

---

[30] Ibid., 600, 618–19.
[31] Hooker, *Laws of Ecclesiastical Polity*, 1:328–66.

common sort of men persuaded, that the wisest, the godliest and the best learned in all Christendom are that way given. . . . [H]ow cometh it to pass they cannot abide that authority should be alleged on the other side, if there be no force at all in authorities on one side or other.

Hooker explains it is no longer necessary to appeal every question of law to divine law, for human authority is sufficient to stand in determinations of justice. As *measures*, laws may be readjusted to new applications: "[L]aws are instruments to rule by, and . . . instruments are not only to be framed according unto the general end for which they are provided, but even according unto that very particular, which riseth out of the matter whereon they have to work."[32]

The trouble is that Hooker's rationale for reform still begs the question of legal criteria. He challenges his adversaries: "They should not therefore urge against us places that seem to forbid change, but rather . . . set down some measure of alteration, which measure if we have exceeded, then might they therewith charge us justly." The problem once again, as it was from the beginning, is how men might agree on a common, fitting "measure of alteration," beyond which change, by consensus, is deemed too much. No wonder that Hooker concedes that the greatest difficulty is not so much applying the law as appraising and approving it: "Easier a great deal it is for men by law to be taught what they ought to do, than instructed how to judge as they should do of law." For all his certainties, Hooker understands that judging the law—given an infinitely deferred, definitive human "measure"—is a formidable task, perhaps the most difficult of all human endeavors: "Soundly to judge of a law is the weightiest thing which any man can take upon him."[33]

Although there were already signs of stress in the relationship between "law" and "truth" in the sixteenth century, Thomas Hobbes is generally credited as the first legal philosopher to propose a clean divorce between them. In his *Dialogue between a Philosopher and a Student of the Common Laws of England* (published posthumously in 1681) and in his *Leviathan* (1651), Hobbes locates the source of law not in God, reason,

---

[32] Ibid., 328, 384.
[33] Ibid., 403, 278.

or nature but the state: "It is not wisdom, but authority that makes a law."[34] Reversing the earlier notion that "an unjust law is no law," Hobbes's positivism conflates justice and legality, rather than justice and truth, as one recent philosopher of law explains it: "[For Hobbes], [t]o ask if a law is just is merely to ask if it is legal. The whole problem reduces itself to a tautology."[35] But although Hobbes's dissociation of law and justice (in any absolute or essential sense) is often held to be the origin of modern jurisprudence, it is essential to recognize the extent to which earlier legal theorists had already confronted the problem of "norms" in human law. Francis Bacon was among those who foresaw a crisis and considered what was necessary to avert it: "There is a strange and extreme difference in laws; some being excellent, some moderately good, and others entirely vicious. I will therefore set down, according to the best of my judgment, what may be called [a] certain 'law of laws,' whereby we may derive information as to the good or ill set down and determined in every law."[36] What some sixteenth-century writers, Shakespeare among them, already understood, was that the formulation of a "law of laws" was as problematic as the formulation of the laws themselves; the effort to create a "metanormative" standard only deferred the crisis to another, human law.[37] Although jurists such as Edward Plowden may have defined the legal process as a progress from "[i]ncertaintie at the beginning . . . reduced to certaintie,"[38] Shakespeare, for one, remained unsure: at the hands of human judges, at least, there may be no "law of laws" to resolve the uncertainties of legal measurement.

If the grounds of law, the ultimate "measure" of legal measures, were vulnerable to skepticism in the Renaissance, so too was the use of law, the application of particular rules to human acts. The ancients and their medieval and Renaissance heirs continually described justice as the discov-

[34] Thomas Hobbes, *Dialogue between a Philosopher and a Student of the Common Laws of England*, in *The Collected Works of Thomas Hobbes*, 10 vols., ed. Sir William Mollesworth (Bristol: Thoemmes Press, 1994; reprint of 1840 ed.), 5.
[35] John A. Alford, "Literature and Law in Medieval England," *PMLA* 92 (1977): 941–51, quote at 948.
[36] Francis Bacon, *The Advancement of Learning*, in *The Works of Francis Bacon*, vol. 3., ed. James Spedding (New York: Garret Press, 1968), 89.
[37] Anna Pintore, *Law without Truth* (Liverpool: Deborah Charles Publications, 2000), 226.
[38] Edward Plowden, *An Exact Abridgment in English of the Commentaries, or Reports of the learned and famous Lawyer, Edmond Plowden* (London, 1650), 4.

ery, or creation of, "equalities"; correspondingly, the result of a just application of the law was called the "equal." As Aristotle explained in his *Ethics*, "the unjust is the unequal, the just is the equal—a view that commends itself to all without proof."[39] Following Aristotle, Aquinas explained that the "right" or the "just" is defined by the commensuration of one person with another, specifically, in a relation of equality: "[T]he term 'justice' denotes equality. . . . [J]ustice is by its very nature concerned with the relation of one thing to another. For nothing is equal to itself, but only to something else."[40] Throughout Renaissance writing, the word "equality" is identified not only as an attribute of justice but as its sine qua non. Shakespeare is representative here: Berowne in his wisdom avers that "justice always whirls in equal measure" (*Love's Labor's Lost*, 4.3.381); Saturninus desires "the extent / Of egall justice" (*Titus Andronicus*, 4.4.3–4); Henry VI promises to "poise the cause in justice' equal scales" (*2 Henry VI*, 2.1.200); while the Archbishop of York takes his own measure of the wars, "I have in equal balance justly weigh'd / What wrongs our arms may do, what wrongs we suffer / And find our griefs heavier than our offenses" (*2 Henry IV*, 4.1.67–69).

"Equality," in other words, is often just another way of saying "justice" in early modern English. The same is true of the word "equity." As noted earlier, "equity" often referred to a special kind of justice, one involving the redress or suspension of the law, but the word was also used to denote "justice" in a general and universal sense.[41] In other words, although they were distinguished by variant meanings, both "equality" and "equity" were synonyms for justice in the Renaissance. Their synonymy had its roots in Latin: both *aequalitas* and *aequitas* were derivations of *aequus*, "equal," and both referred in Latin to "evenness," "levelness," "proportion," "balance," "fairness," or "justice." In Renaissance English, both continued to be understood as versions of *aequus*—as Aquinas stated, the results of "a right or just act . . . which is 'adjusted' to someone or something according to *some kind of equality*."[42] The rhetorical confusion between these terms speaks to a philosophical quandary of the period: which *kind* of "equality" was the most just, the most "equal"?

Spenser's legend of Justice in the *Faerie Queene* considers some of

[39] Aristotle, *Nichomachean Ethics*, V.1.7.
[40] Aquinas, *Summa theologiae*, IIaeIIae 58, p. 172; 57, p. 159.
[41] See the *Oxford English Dictionary*, s.v. "equity."
[42] Aquinas, *Summa theologiae*, IIaeIIae 57, p. 161, emphasis added.

equality's "kinds." Spenser's "champion of true Iustice" is "Artegall."[43]
His name is routinely glossed as signifying "equal to Arthur," yet it is
likely that Spenser also means to identify him with the "art" of making
"equal," as the practice of justice was commonly understood in the period. Spenser relates how Astraea, having abandoned a world "runne
quite out of square" (5.Proem.1.7) and departed the earth, instructs
men in the "rules of justice" (5.1.5.9), making Artegall its "instrument"
(5.Proem.11.9). Astraea gives him a set of scales:

> There she taught him to weigh both right and wrong
> In equall balance with due recompence.
> (5.1.7.1–2)

Opposed to Artegall is a Giant who boasts "[t]hat all the world *he* would
weigh equallie" (5.2.30.5, emphasis added) with his own set of scales:

> He sayd that he would all the earth vptake,
> And all the sea, deuided eache from either:
> So would he of the fire one ballaunce make,
> And of th'ayre . . .
> .   .   .   .
> For why, he sayd, they all vnequall were.
> (5.2.31.1–4, 32.1)

Artegall condemns the Giant's "egalitarian" measures:

> For take thy ballaunce, if thou be so wise,
> And weigh the winde, that vnder heauen doth blow;
> Or weigh the light, that in the East doth rise;
> Or weigh the thought, that from mans mind doth flow.
> But if the weight of these thou canst not show,
> Weigh but one word which from thy lips doth fall.
> (5.2.43.1–6)

Alluding to Solomon's Wisdom and to Isaiah ("Who hath measured the
waters in his fist and counted heaven with the span, and comprehended

---

[43] Edmund Spenser, *The Faerie Queene*, ed. Thomas P. Roche Jr. (New Haven: Yale
University Press, 1978), 5.1.32. Further quotations from the *Faerie Queene* are drawn
from this edition and appear parenthetically in the text.

the dust of the earth in a measure? And weighed the mountains in a weight, and the hills in a balance?" [40:12]), Artegall reminds the Giant that the elements of earth

> [A]t the first they all created were
> In goodly measure, by their Makers might,
> And weighed out in ballaunces so nere,
> That not a dram was missing of their right.
> (5.2.35.1–4)

In relation to God's measures, men, even giant men, are unequal to the art of justice.

When the Giant, in shame, moves to destroy his scales, however, Artegall stops him: "Be not vpon thy balance wroken / For they doe nought but right and wrong betoken" (5.2.47.4–5).[44] According to Artegall, the giant's scales make literal what is better apprehended by reason and discretion: "But in the mind the doome of right must bee. . . . The eare must be the balance, to decree / And iudge" (5.2.47.6, 8–9). Artegall identifies the scales of the mind with "equitie," which "measure[s] out along, / According to the line of conscience" (5.1.7.3–4). Like many of his contemporaries, Spenser privileges the moral and mental "spirit" of measurement over its "letter," and equity emerges as the highest *kind* of equality—*more* "equal" than others.

If conscience or mind is the measure, Spenser's equality is justifiably deemed an "art," understood as an acquired, human skill, something opposed to nature. Yet Plato, in his *Laws*, insisted that "equality" was an objective determination, a quality that inheres in the relations of things and not in minds:

> The reason why the equal is equal, or the symmetrical symmetrical, is not at all because a man so opines, or is charmed thereby, but most of all because of truth, and least of all for any other reason.[45]

---

[44] For further critical discussions of this episode, see Annabel Patterson, "The Egalitarian Giant: Representations of Justice in History/Literature, *Journal of British Studies* 31 (April 1992): 97–132; and Elizabeth Fowler, "The Failure of Moral Philosophy in the Work of Spenser," *Representations* 51 (Summer 1995): 47–76.

[45] Plato, *Laws*, 667E–668A.

The purpose of law, in fact, is to guarantee that a single truth will apply in all determinations of equality. Aquinas, citing Aristotle, observed that the existence of laws ensures that truth, not opinion, governs the administration of justice:

> The Philosopher says at Rhetoric I, "it is better that all things be regulated by law than left to the decision of judges." . . . Since, then, the "animate justice" of the judge is not found in many men, and because it can be distorted, it was therefore necessary, whenever possible, for the law to determine what the judgment should be, and for very few matters to be entrusted to the decision of men.[46]

Hooker concurred, explaining that the first lawgivers "knew that no man might in reason take upon him to determine his own right, and according to his own determination proceed in maintenance thereof, inasmuch as every man is towards himself and them whom he greatly affecteth partial." And although citizens may not heed an individual's judgment against them

> when they are told the same by a law, think very well and reasonably of it. For why? They presume that the law doth speak with all indifferency; that the law hath no side-respect to the persons; that the law is as it were an oracle proceeded from wisdom and understanding.[47]

Shakespeare makes continual reference to the oath that Renaissance monarchs took at their coronation, to provide "equal and right justice between party and party," despite their personal allegiances.[48] For example, King Richard II promises an "equal" hearing to Mowbray and Bullingbroke ("impartial are our eyes and ears" [1.1.115]), while John of Gaunt sentences his own son to avoid "a partial slander" (an imputation of partiality [1.3.241]) in his judgment.

---

[46] Aquinas, *Summa theologiae*, IaIIae 95, pp. 128–29.
[47] Hooker, *Laws of Ecclesiastical Polity*, 1:242, 244.
[48] James S. Hart Jr., *The Rule of Law, 1603–1660: Crowns, Courts, and Judges* (Harlow: Pearson Longman, 2003), 21.

What Shakespeare invokes in these passages is the principle of the "rule of law"—the doctrine, ultimately derived from natural law theory, that "in order to control the exercise of arbitrary power, the latter must be subordinated to impartial and well-defined principles of law."[49] As an instrument of justice, the rule of law applies the same measure across cases; it functions, in other words, as a *standard* measure. As legal historians have noted, the ideal of a common, constant, and consistent application of the law was fundamental to an early modern understanding of jurisprudence:

> The cultural ethos of early modern England was shaped by many common perceptions and beliefs, but none more powerful than the universal faith in the rule of law. . . . Recourse to "the law" had become almost an instinctual response to conflict, at all levels and in all cases. The law's reach was meant to be universal and impartial, its protections available (and its sanctions applicable) to one and all.[50]

The impartiality of law is one crucial way in which justice, then as now, lays claim to the equality of its operations:

> If the justice in all kinds of human transactions is to be measured effectively, those transactions have to be governed by rules which are applied with as much consistency as it is possible to achieve. What this requires is the formalization, and hence the depersonalization, of justice. . . . Moral principles and standards have to be formalized into unbending rules which apply to the act, rather than the actor. When judicial independence is established, the ideal of impartiality—itself a precondition of equality before the law—can be developed.[51]

The trouble, once again, is that there is more than one kind of "equality under the law" recognized by Renaissance legal theorists. "Eq-

---

[49] *Oxford English Dictionary*, s.v. "rule of law." The phrase "rule of law" is first used by Edward Coke in 1644, in reference to a "valid legal proposition"; its modern sense arises in the nineteenth century.
[50] Hart, *Rule of Law*, 1.
[51] Tebbit, *Philosophy of Law*, 8.

uity" was "equal" too. In its identification with natural law, "equity" was held to provide the standard measure: "We must urge that the principles of equity are permanent and changeless, and that the universal law does not change either, for it is the law of nature, whereas written laws often do change."[52] Equity insists that laws remain equal to or commensurate with the acts to which they are applied, as Aquinas wrote: "[A] law is framed as a rule or measure of human acts. Now a measure should be in keeping with what it measures, as stated at *Metaphysics X*, for different things are measured by different measures. Hence the laws imposed on men should also be in keeping with their condition."[53] Hooker concurred: "For inasmuch as the hand of justice must distribute to *every particular* what is due, and judge what is due with respect no less of particular circumstances than of general rules and axioms, it cannot fit all sorts with one measure, the wills, counsels, qualities and states of men being divers."[54] Once again, the opposition between law and equity, along with their common claims to "equality," raised that difficult question: Which was more "equal"—a law that applies equally (in the same way) to all, or one that is equal to the differences of men?

In the Renaissance, *both* purported to provide a universal measure based on a general criterion of "reason," "nature," or divine sanction. Hobbes's Philosopher would challenge both in turn, arguing that, in either case, there is no such thing as a standard judgment among divers men. Regarding equity, he states that "[t]here is not amongst men a universal reason agreed upon in any nation, besides the reason of him that hath the sovereign power. Yet though his reason be but the reason of one man, yet it is set up to supply the place of that universal reason." For Hobbes, both the law and its suspension represent no more and no less than the will of a single man who rules: "A law is the command of him or them that have the sovereign power, given to those that be his or their subjects, declaring publicly and plainly what every of them may do, and what they must forbear to do."[55] Shakespeare often identified the extremity of the law with the whim of a harsh judge; consider Hermia's death sentence at the hands of the Duke and her father at the start of *A Midsummer Night's Dream*, or Lear's disinheritance of Cordelia, to take

[52] Aristotle, *Rhetorica*, quoted in Tebbit, *Philosophy of Law*, 13.
[53] Aquinas, *Summa theologiae*, IaeIIae 96, p. 140.
[54] Hooker, *Laws of Ecclesiastical Polity*, 2:514–15.
[55] Hobbes, *Dialogue*, 22, 26.

just two examples. What is far more surprising, however, is how often Shakespeare, anticipating Hobbes, exposes equity too as another expression of private will posing as "true," "common," or "natural" justice. In his concerns about the "measure" of equity, Shakespeare seems to have had a lot in common, once again, with John Selden, who wittily described the defects of equity's "rule":

> Equity is a Roguish thing, for Law we have a measure, know what to trust to, Equity is according to the Conscience of him that is Chancellor, and as that is larger or narrower, so is Equity. 'Tis all one as if they should make the Standard for the measure, we call [a Foot] a Chancellor's Foot; what an uncertain Measure would this be! One Chancellor has a long Foot, another a short Foot, a Third an indifferent Foot: 'Tis the same thing in the Chancellor's Conscience.[56]

By comparing "equity" to "anthropometric" measures (see chapter 1) that embody only a particular man's physical dimensions, Selden asserts "law" to be the more certain standard.[57]

The Court of Chancery, the independent Renaissance court of equity established to serve as a check and a balance to common law, was officially held to represent the king's "most wise discretion," his "acting conscience."[58] As Shakespeare would show again and again, however, the standard of the king's conscience was hardly a guarantor of justice. Hobbes's Lawyer reasoned that there ought to be no need for separate courts: "Seeing all judges in all courts ought to judge according to equity, which is the law of reason, a distinct court of equity seemeth to me to be unnecessary, and but a burthen to the people, since common-law and equity are the same law." His Philosopher, in response, makes a crucial qualification, and one long since anticipated by Shakespeare: "It were so

---

[56] John Selden, *The Table-Talk of John Selden* (London: William Pickering, 1847), 64.
[57] Selden's French contemporary, however, Jean Bodin, in his *Six Bookes of the Commonweale* (1606), used the idea of anthropometric measures to prove, on the contrary, that laws must be multiform: "[W]hat Shoo maker is so ignorant or foolish, as to shape one fashioned shoo, or of the same last, to euery man's foot?" See Bodin, *Six Bookes of the Commonweale*, facsimile ed., ed. Kenneth Douglas McRae (Cambridge, MA: Harvard University Press, 1962), 781.
[58] Hart, *Rule of Law*, 26.

indeed, if judges could not err."[59] Hooker too had already observed that the administration of equity, no less than the law itself, might be exploited for personal advantage, especially when the judge himself has a stake in the case: "[B]ecause it is natural unto all men to wish their own extraordinary benefit, when they think they have reasonable inducements so to do . . . no man can be presumed a competent judge [of] what equity doth require in his own case."[60] As a procedure for justice, equity is most doubtful, it seems, in the case of self-judgment.

I turn now to Shakespeare's *Measure for Measure*, a play that examines the problem of partiality in legal and literary judgments. Shakespeare's last comedy, like so many before it, begins with a threatened imposition of an overly harsh, strict, or inflexible rule and ends with a qualification of that rule. In Christian terms, *Measure for Measure* traces a shift from Old Law to New. The "old law," in this case, is one that bans fornication outside of marriage and sentences violators to death. The Duke's relationship to this law, which he suspends, then revives, and then suspends all over again, has often been challenged, including his private motives in determining the course of justice. There can be no doubt that the play condemns the Duke's magistrate, Angelo, who enforces the law only to breach it himself. Yet contemporary readings of *Measure for Measure* have not gone far enough in assessing the consequences of Shakespeare's skepticism toward legal judgment. The play may proceed from law to equity, yet Shakespeare sees no progress here; the end only reinstates the philosophical problem of judgment with which it begins. The title of the play, *Measure for Measure*, ultimately points to an impasse in the advance of legal reform: it characterizes the story in terms of an exchange of measures, one measure for another, that is no exchange at all. In *Measure for Measure*, neither the law nor its "corrections," such as equity, represent fixed or universal standards but rather the changing wills of the judges who apply them. If the play begins with one uncertain "measure," it ends with another that is as contingent and corruptible as the first.

A key to understanding Shakespeare's critique of legal judgment appears in a brief exchange between Escalus, the wise elder statesman, and Pompey, the unrepentant pimp,

---

[59] Hobbes, *Dialogue*, 26.
[60] Hooker, *Laws of Ecclesiastical Polity*, 2:39.

ESCALUS: How would you live, Pompey? by being a bawd? What do
   you think of the trade, Pompey? is it a lawful trade?
POMPEY: If the law would allow it, sir.
(2.1.224–27)

When Escalus asks Pompey what he thinks about being a bawd, whether
he considers it "lawful," he implicitly identifies lawfulness with morality.
Although it is Escalus to whom the Duke attributes the highest knowledge of "the terms / For common justice" in Vienna (1.1.10–11), Pompey comes closer to the crux of the matter: he answers the question in
Hobbesian terms, as if law were a matter of legality only. Crucially, his
tautological response (pimping would be lawful if the law allowed it) is
neither simple nor sophistical in the context of Shakespeare's examination of the legitimacy of the law. Young Claudio, sentenced to die under
the newly enforced laws, similarly demystifies the relationship between
"authority" and what is, or what is considered to be, right: "[O]n whom
it ["Authority"] will, it will; / On whom it will not, so; yet still 'tis just"
(1.2.122–23). When Pompey is "rehabilitated" and given a job as a hangman (he begins the play as a forfeit to the law and ends up imposing that
forfeit on others), he remarks, "I have been an unlawful bawd time out of
mind, but yet I will be content to be a lawful hangman" (4.2.15–16). If
the irony of being a "lawful" executioner is unintentional on Pompey's
part, the Provost makes it clear that there will be no moral gain in the exchange. He measures the bawd and the hangman thus: "[Y]ou weigh
equally; a feather will turn the scale" (4.2.30–31). In a world in which the
meaning of "lawfulness" is continually under investigation, Escalus himself is compelled to challenge the distinction between the judge and the
judged, with a question that will haunt the play as a whole, "Which is the
wiser here: Justice or Iniquity?" (2.1.172).

Measure for Measure is a play in which, one by one, those who stand in
judgment over others end up perpetrating the very crimes they set out to
punish. Angelo attempts to compel Isabella to "lay down the treasures of
[her] body" for him in exchange for saving her brother from the rigors
of the law. The Duke arranges a "bed trick" between Angelo and Mari-
anna, even as he condemns Pompey for the "filthy vice" of being a
"whoremonger" (3.2.23, 36). Both Angelo and the Duke originally present themselves as men who do not "detect" for women (3.2.121); both
end up pursuing Isabella, despite or perhaps because of her determination to take vows of chastity and become a nun. All this is clear enough

and long rehearsed in scholarly criticism of the play. Yet it would be a mistake to draw from this what might seem the simplest and most obvious conclusion—that *Measure for Measure* is, fundamentally, an indictment of hypocrisy. Shakespeare's study of justice is not concerned, finally, with problems of moral degeneracy, duplicity, or double-dealing. Rather, the play suspends judgment of what is right and wrong to address the a priori, skeptical problem of how "rightness" or "wrongness" is determined—by the "measures" of judgment itself.

In the first scene of the play, Duke Vincentio admits that he is to blame for the sexual license of his citizens: "'[T]was my fault to give the people scope" (1.3.35). It seems he seeks to have that scope rescinded or at least reduced. When, however, he makes Angelo his deputy, he does not tell him to limit the people's scope; in fact, he does not charge Angelo with any prescribed rule. Rather, he instructs him, "Your scope is as mine own, / So to enforce or qualify the laws / As to your own soul seems good" (1.1.64–66); the reach of the law, he insists, depends solely on what his substitute deems "good." Implicitly, the people's scope, then as now, depends on and reflects that of their judge. Angelo has a reputation for "straightness" (Escalus says he believes him to be "most strait in virtue" [2.1.9]; the Duke commends the "straitness of his [Angelo's] proceeding[s]" [3.2.255–56]), but he, no less than the Duke, has the authority to bend laws at will. The Duke's understanding of the relationship between ruler and rule seems to anticipate that of Jean Bodin, the first early modern political philosopher to identify political power with the authority, above all else, to make laws or change them: "Wherefore let this be the first and chiefe marke of a soueraigne prince, to bee of power to giue lawes to all his subiects in generall, and to euerie one of them in particular. . . . without consent of any other greater, equall, or lesser than himselfe."[61] As sovereign, the Duke and his chosen deputy measure their subjects by the "full line of [their] authority" (1.4.56).

There is, however, a catch to having a full line of authority over law. As every scholarly edition of *Measure for Measure* notes, the title of the play invokes Jesus's injunction, in Matthew 7:1, "Judge not, that ye be not judged. For with what judgment ye judge, ye shall be judged; and with what measure ye mete, it shall be measured to you again." Aquinas glossed the passage this way: "Those that are guilty of grievous sins should not judge those guilty of the same or lesser sins"; "If we find that

---

[61] Bodin, *Sixe Bookes*, 159.

we ourselves are guilty of the same sin as someone else, let us deplore the fact together with him, and invite him to join with us in striving against it."[62] In his efforts to move Angelo to mercy toward Claudio, Escalus implores the deputy to examine himself, to consider whether he might be guilty of the same illicit (or illegal) desires as the man he condemns to die:

> Let but your honor know
> (Whom I believe to be most strait in virtue)
> That in the working of your own affections,
> Had time coher'd with place, or place with wishing,
> Or that the resolute acting of [your] blood
> Could have attain'd th'effect of your own purpose,
> Whether you had not sometime in your life
> Err'd in this point which now you censure him,
> And pull'd the law upon you.
> (2.1.8–16)

Angelo assures him in kind,

> When I, that censure him, do so offend,
> Let mine own judgment pattern out my death,
> And nothing come in partial.
> (2.1.29–31)

Isabella follows the same strategy in pleading for mercy on behalf of her brother: "[A]sk your heart what it doth know / That's like my brother's fault" (2.2.137–38). At the end of the play, knowing full well Angelo's treachery, the Duke expresses incredulity that Angelo would sentence Claudio for a vice that they share:

> it imports no reason
> That with such vehemency he should pursue
> Faults proper to himself. If he had so offended,
> He would have weigh'd thy brother by himself,
> And not have cut him off.
> (5.1.108–12)

---

[62] Aquinas, *Summa theologiae*, IIaeIIae 60, pp. 195–96.

Shakespeare understands the injunction "Judge not, that ye be not judged" as a behest to "weigh" one's brother by oneself, to make one's moral self the measure of others'. Justice is a scale in which the judge's vices must be compared to those whom he judges; the judge must proceed by neither "more nor less to others paying / Than by self-offenses weighing" (3.2.265–66). Yet taking the logic of Matthew 7:1 to its limit, the Duke goes further: he maintains that Angelo's judgment of Claudio was just, and would have remained just, as long as Angelo himself had remained clear of the same offense. The Duke makes this explicit: "If his [Angelo's] own life answers the straitness of his proceeding it shall become him well; wherein if he chance to fail, he hath sentenc'd himself" (3.2.255–57). Claudio accepts Angelo's sentence as just, as the Duke (disguised as the Friar) confirms to Escalus: "He [Claudio] professes to have receiv'd no sinister measure from his judge, but most willingly humbles himself to the determination of justice" (3.2.242–44). It's not that the old laws against fornication are in themselves unfair. Or more precisely, there is no way to make an independent evaluation of the law so long as one man, one judge, is the measure. All we can know is this: as long as Angelo is "strait," the law is "right."

Isabella is often compared to Angelo, considered his "equal," at least, in the way that she too asserts an initial allegiance to a very "strait" law. She is introduced to us in act 1, scene 4, as a woman whose chief desire is "a more strict restraint" on her liberties than even, perhaps, the rigorous order of St. Clare will demand of her. Isabella's case, however, is different than Angelo's. Although he claims to hold himself to the *same* standard to which he holds others, she declares herself subject to a "higher" law. In her suit to save her brother's life, Isabella is prepared to appeal to a commonality of vices among others; indeed, she urges Angelo to see a parallel between her brother's fault and his own human fallibility. Yet when the treacherous magistrate asks her to do the same (he argues that her sin in having sex with him would be no greater than her brother's sin), she abjures the principle of human equality, insisting, as she had remarked to Angelo earlier, that "[w]e cannot weigh our brother with ourself" (2.2.127). When she agrees to the Duke's plan to send Angelo's jilted lover, Mariana, to his bed in her place, she does not compare herself to her: just as she places her body and soul above Claudio's ("More than our brother is our chastity" [2.4.185]), she judges Mariana's as "less" than her own. Does Shakespeare consider Isabella's observance of the law of the convent—arguably a "divine" law—a legitimization of her judgment of others? Probably not. Isabella is only, after all, an initiate. For the moment, she proves as human as the rest of them, as she (in all likelihood) chooses to exchange her promise of marriage vows to God for marriage vows to the Duke. He alone has always been, and will remain, her law.

When Isabella makes her entrance in the play, she leads with a question, "And have you nuns no farther privileges?" (1.4.1). By "privileges" Isabella appears to mean "liberties" or license in a general sense. But "privilege" is also a legal term in early modern English; it means, literally, "private law." Thomas Elyot's mid-sixteenth-century *Dictionary* defines *priuilegium* as "a law concernynge priuate persones. Also a priuate or speciall lawe."[63] As Aquinas had explained, "[T]here are some laws which affect the community in one respect and the individual in another. These are called 'privileges' [privilegia]: 'private laws' [leges priuatae] as it were, because they have to do with individual persons."[64] Bodin explains further that "[a] priuilege I call a law made for one, or some few particular men: whether it bee for the profit or disprofit of him or them for whome it is graunted."[65] Although "privilege" denoted a proper category within the common law, it often connoted special pleading and unfair advantage. Hooker defined it: "A privilege is said to be that, that for favour of certain persons cometh forth *against* common right."[66] Richard II identifies "privilege" with "partiality" when he promises to give his cousin Bullingbrook a fair hearing: "Such neighbor nearness to our sacred blood / Should nothing privilege him nor partialize / The unstooping firmness of my upright soul" (1.1.119–21). There is often a suggestion of class prerogative in uses of the word, as when Prince Hal is warned, "[T]hou hast lost thy princely privilege / With vile participation" (*1 Henry IV*, 3.2.86–87). Above all, privilege is the law of the individual, of the "self," in contradistinction to laws that govern all others, the "common" law: "Be where you list, your charter is so strong, / That you yourself may privilege your time / To what you will, to you it doth belong / Your self to pardon of self-doing crime" (sonnet 58, lines 9–12).

---

[63] Thomas Elyot, *The Dictionary* (London, n.d.), s.v. "priuilegium."
[64] Aquinas, *Summa theologiae*, IaIIae 96, p. 137.
[65] Bodin, *Sixe Bookes*, 160.
[66] Hooker, *Laws of Ecclesiastical Polity*, 2:512 (emphasis added).

According to Bodin, the source of privilege, like the source of all law, can only be the monarch or the "sovereign" self; in the political sphere, all "priuileges . . . only belong vnto soueraigne princes to graunt."[67] The "private order" (5.1.466) that saves Claudio from execution is just one of the "privileges" Duke Vincentio extends to his people. Indeed, each of the Duke's final dispensations of justice represents an exceptional ruling, granted on a case-by-case basis; all are in direct conflict with the rulings of the law. All law in *Measure for Measure* is exposed as "private" law, the privilege of the ruler. This is the case for the common laws of Vienna, but also, crucially, for its qualifications: equity too is the "privilege" of the Duke. In a world in which one judge varies from the next, or in which judges themselves may change, what constitutes justice, what is deemed legal and what is not, changes as well. And in such a world, the difference between law and equity ultimately is abolished: call it "law" or call it "equity," justice in *Measure for Measure*, from beginning to end, is at the sovereign's discretion. It conforms not to the contours of the law, or even to individual cases in law, but to the Duke's own cause of seeing "[his] pleasure herein executed" (5.1.521).

But what is the Duke's pleasure, precisely, in the case of Vienna? He is notoriously mysterious about it. He says that it has something to do with reinstating the olds laws of the city; he states that he wants to test Angelo's "straitness" under the pressures of absolute power. He adds that there are "[m]oe reasons for this action" (1.3.48), which he hasn't time or will to divulge. We may speculate that he means to right the wrongs done to Mariana or, depending on how far one imagines his omniscience extends, to marry Isabella. Perhaps his chief, undisclosed reason, one implicit in the course of his proceedings, is his desire to put on trial the justice of *his* once, and future, administration. Angelo's "straight" rule is a standard by which to measure his own bending one; Angelo's success would have meant, by comparison, the Duke's failure as a magistrate. Angelo's failure, in turn (that is, his failure to stay "straight"), suggests that the Duke was right all along to give the people "scope." After all, Angelo's fall confirms the play's repeated claims about human equality, the commonality of our frailties and fallibilities. Angelo's fall serves, above all, to justify the Duke's own faults. For the Duke, the cause of common justice is finally less consequential than the cause of self-justification.

---

[67] Bodin, *Six Bookes*, 45.

Clearly, the Duke was never one to take criticism, from any dark corner, kindly. He admits to Friar Thomas a fear of slander (1.3.43) and passes what is arguably his harshest judgment on Lucio, whom he claims has continually slandered him during his absence from Vienna. Crucially, what the Duke calls slander is, often enough, Lucio's imputation that the Duke is "equal by nature" to his subjects. When Angelo sentences Claudio to die, for example, Lucio asks, "Would the Duke that is absent have done this?" (3.2.116–17). The answer, as we know as well as Lucio does, is certainly not: the Duke has never and will not sentence a young man who has had sex with his betrothed to die (we have the Duke's "rescue" of both Claudio and Angelo as further proof). Lucio explains why the Duke had always granted mercy toward fornicators: "He had some feeling of the sport; he knew the service, and that instructed him to mercy" (3.2.119–20). Lucio interprets the Duke's lenity as an expression of fellow feeling, an admission of his own guilt, exactly as the Duke sees Angelo's: "When vice makes mercy, mercy's so extended, / That for the fault's love is th'offender friended" (4.2.113–14).

Although the Duke may deny the correlation, *Measure for Measure* makes it clear that for human judges, at least, understanding our shared moral failings is the very basis of mercy (for Shakespeare, only God's mercy is not based on the equality between judge and judged). We have no particular evidence that the Duke had any "feeling of the sport" until he met Isabella. (He explicitly denies any interest in women to Friar Thomas, but then so does Angelo, who "scarce confesses / That his blood flows" [1.3.51–52].) We do know, however, that the Duke is ready to help others to the same "fault." The Duke pronounces his verdict on Angelo's villainy: "Twice treble shame on Angelo / To weed my vice and let his grow" (3.2.269–70). When he refers to "my vice," the Duke may mean the vice of his people, conceived in metonymical relation to himself, but his phrasing is surely revealing: he knows that Angelo, implicitly, is judge of his vices, especially the Duke's own "license." Vincentio's early comment to Friar Thomas must be understood as a confession: "'Twas my fault to give the people scope" (1.3.35). His leniency answers for his people's laxness, his lack of rule their misrule. As his own insistence on the need for "equality under the law" requires, the Duke's slackness set the standard for his people's. As he himself puts it, "Thieves for their robbery have authority / When judges steal themselves" (2.2.175–76). They had no choice but to "[b]e *ruled* by him" (4.6.4, emphasis added).

Indeed, *Measure for Measure* does not, finally, ascribe the problem of "misrule" to the crimes and misdemeanors of Vienna's lawbreakers. Shakespeare seems to have little stake in teaching a lesson to Claudio or even to Pompey, neither of whom, arguably, has undergone a moral transformation as a result of their brush with the law. Shakespearean "misrule" appertains to the fault of judges rather than to those judged, the "mismeasure" of laws and those who apply them. No doubt the old law was at times too broad in its application (under Angelo's rule there was no way to distinguish "degrees of guilt" among the offenders, Claudio, Pompey, and Mistress Overdone) or underextended (we are told that while the brothels in the suburbs are all torn down, a "wise burgher" manages to keep the brothels in the city up and running [1.2.100]). The Duke's equitable rule, however, is no "fitter." His indiscriminate mercy, for example, makes no moral distinction between Claudio and Barnardine, the murderer, when even the Provost, who often seems the most reliable witness to the play's proceedings, sees the difference: "Th'one has my pity; not a jot the other, / Being a murtherer, though he were my brother" (4.2.61–62). In this as in many of his judgments, the Duke arguably remains, as Lucio puts it, "[a] very superficial, ignorant, *unweighing* fellow" (3.2.139–40, emphasis added).

It is likely that "misrule" is inevitable, according to Shakespeare, in any attempt to "measure" human sexuality. As many scholars have suggested, the playwright may well concur with the bawd Pompey, who argues that the only effective laws against sex would be ones that dismembered the human body (2.1.230); laws will always exceed or fall short of the scope of human desire, however "natural" the laws purport to be (i.e., based in reason, conscience, or divine injunction). *Measure for Measure* intimates that there is no such thing as a "normal" sexuality, in the sense that human sexual behavior and sexual feeling will not subject itself to any "standard" procedures. The "ruler's" sexuality, implicitly, creates rather than discovers what will be deemed "normal" desires, whether or not they are shared by others.

Duke Vincentio, in any case, never says anything definitive regarding how natural, how good, or how wise are his own original injunctions against fornication. As in so many of Shakespeare's comedies, which begin with the threatened imposition of a harsh sentence, the status of the law is left uncertain at the end of the play. In place of such certainty, Shakespeare can only offer this: "You will think you have made no offense, if the Duke avouch the justice of your dealing" (4.2.185–86). If anyone should challenge his verdicts, the Duke can always borrow Angelo's retort to Isabella: "As for you, / Say what you can: my false *o'er-weighs* your true" (2.4.169–70, emphasis added). It is no wonder that scholars have so often identified Shakespeare's Duke Vincentio as a figure for the artist or the playwright who orchestrates the Venetian scene to his own liking. Whether or not he is the most gifted or the most ethical of Shakespeare's artists, he makes his own law, sets his own standards, with the best of them. Art, for Shakespeare, is its own privilege, its own "ruling" desire—all well and good, until art is mistaken for truth, one man's pleasure for every man's law.

There's no question that in many of his plays Shakespeare characterizes a strict observance of the letter of the law as a sign of an indifference to human nature, even of *inhumanity*. Angelo joins villains such as Shylock, Solinus, and Theseus, who "stand for law" at the expense of human life. It may thus seem counterintuitive, if not perverse, to suggest that Shakespeare renders the extenuation of the law as, potentially, a comparable abuse. Duke Vincentio's personal motives have met with a great deal of criticism in modern readings of the play, yet "equity" itself, it seems, has remained unimpeachable. The problem may be our long-standing if implicit critical faith in a Christian teleology that celebrates the supersession of absolute justice by a principle undeniably more complex, more nuanced, more sensitive to human frailty and human difference. Recent studies in law and literature betray the same bias. Kathy Eden, notably, writes persuasively of the imaginative as well as the ethical superiority of equity over law, as conceived by classical writers and their heirs. These authors, she suggests, implicitly ally equity with literature itself, in their mutual commitment to "render[] the individuality of experience more demonstrable and therefore more knowable" and in their concern for the intentions "behind" the laws and those who break them. For Eden and others, equity is a model for, or analogue to, literary interpretation itself.[68] By implication, when a writer such as Shakespeare invokes the cause of equity, makes it the climax of his dramatic denouements, he is making a plea for his own art.

In his deepening scrutiny of the fallibilities and biases of those who judge, however, Shakespeare ultimately refused to assume that equity

---

[68] Kathy Eden, *Poetic and Legal Fiction in the Aristotelian Tradition* (Princeton: Princeton University Press, 1986). See also Ian Maclean, *Interpretation and Meaning in the Renaissance: The Case of Law* (Cambridge: Cambridge University Press, 1992).

was any more "equal" than the law it aimed to correct. If we imagine that he might have found "equity" to be more "literary" than the law in its sensitivity to the idiosyncrasies of human character, we must allow that he was just as likely to have observed a comparable affinity between the "law" and any general truths, or shared values, he finds among men and women. If he admired an emphasis on interpretation rather than fixed "truths," he did not necessarily judge equitable interpretation a more ethical practice than reading "by the letter." Often enough, Shakespeare represents interpretation, however liberal in its scope, as partial and uncertain, sometimes dangerously so, as his Cicero warns: "[M]en may construe things after their fashion / Clean from the purpose of the things themselves" (*Julius Caesar*, 1.3.34–35). While he may celebrate the artist's freedom to "construe things after their fashion," he is hardly convinced that judgments arrived at in this way are necessarily "right."

As in so many of the playwright's interrogations of the measures of man, Shakespeare reveals the difficulty, if not impossibility, of finding absolute criteria for establishing and certifying the reliability, efficacy, and validity of laws—*and* their interpretations. He concurs with Aristotle that "[t]he rules of justice based on convention and expediency are like standard measures," but he also understood, with Aristotle, that those "standards" inevitably varied according to the contingencies of time and place: "Corn and wine measures are not equal in all places, but are larger in wholesale and smaller in retail markets."[69] Shakespeare, with many others who have written about the legitimacy of human law, perceived that the existence of multiple "standards" potentially undermined the possibility of having standards altogether. The problem with equity, as Aquinas had warned, is that in setting a new standard for every case, the law itself may be invalidated:

> [I]f there were as many rules or measures as there are things measured or ruled, the rules or measures would cease to be of any use, since the usefulness of a rule or measure lies precisely in the fact that it is a single standard which applies to many instances. And so the law would be of no use if it did not extend beyond one single act. For whereas the decrees of men of practical wisdom are given

---

[69] Aristotle, *Nichomachean Ethics*, V.7.5.

for the purpose of directing particular actions, law is a general precept.[70]

A measure that makes too many distinctions makes no distinctions at all: it becomes what Erasmus refers to in his *Adages* as "an unmarked rule."

Erasmus offers three demonstrative illustrations of this adage. Plutarch, in his essay "On Garrulity," explains that "[t]he garrulous person is absolutely an unmarked rule as far as conversation goes, because he talks nonsense with no discrimination about anything whatever." Aulus Gellius, in his *Attic Nights*, used it to describe a method (or perhaps, a nonmethod) of reading: "[A]ll those people, and among them especially the Greeks, read eagerly and widely and swept together everything they came across without discrimination, using a 'blank ruler' as they say, simply intent on quantity alone." Socrates, finally, described his appreciation of beautiful boys as the application of "a white rule against a white stone," "because he did not distinguish clearly between types of beauty, and he loved them all alike." Erasmus concludes that "an unmarked rule," by "making no distinctions," is a metaphor "for those who [whether they would wish it for themselves or no] have no judgment."[71]

Jean Bodin rehearsed the ancient claim that only Lesbian measures do justice to what they measure:

> [T]he Lesbian rule . . . being made of lead, was euerie way so pliant and flexible, as that it might be vnto euery stone so aptly fitted and applied, as that no part (so much as possible was) might thereof be lost: whereas others, who were woont to apply the stone vnto the straight rule, oftentimes lost much thereof.

However, he observes, that is only one opinion:

> So say some, that judges ought in judgement to apply the lawes vnto the causes in question before them, and according to the varietie of the persons, times, & places, so to decline from that inflexible straightnesse. Howbeit in mine opinion as it is impossible

---

[70] Aquinas, *Summa theologiae*, IaIIae 96, p. 137.
[71] Erasmus, *Adagia*, I.v.88.

for a rule so pliant in euery way, as was the Libian [*sic*] Rule, to
keep the name of a rule: so must also the strength and name of the
law perish, which the judge may at his pleasure euery way turne
like a nose of waxe, and so become the master and moderator of
the law, whereof indeed he ought to be but the vpright minister.[72]

What, finally, is Shakespeare's opinion on the matter? Is he equal or
equitable, "straight" or "Lesbian," in his orientation toward judgment?
Shakespeare, we might say, "queers" the law, however straight it purports
to be: for him all law and all interpretations of law are always and already
Lesbian rules, bent or flexible measures that challenge the claims of
human judgment to objectivity or absolute truth. Shakespeare's legal
measures bear a metonymical relation to those who apply them: Man
(often, one particular man) is the measure of what is just. By what criterion, then, can laws themselves be evaluated? In *Measure for Measure*,
Shakespeare tests the proposition, founded on notions of "equality
under the law," that as long as the judge is accountable to his own measures, those measures are just. But this is merely circular: the judge's sub-
jection to the "rule of law" only guarantees that the man who is the measure is consistent with himself. There remains no independent criterion,
no way to know, the value of the laws themselves. At the end of the play, at
the end of his career in comedy, Shakespeare suspends the absolute rule
of law indefinitely and indeterminately and leaves us, instead, at the
mercy of judges—to hope perhaps, but just as likely to new fear.

[72] Bodin, *Sixe Bookes*, 76.

Epilogue

HOW SMART IS HAMLET?

Shakespeare and Renaissance "Intelligence Testing"

THE TITLE OF THIS BOOK is borrowed, of course, from Stephen Jay Gould's *The Mismeasure of Man*, first published in 1981.[1] Gould's title, in turn, is an appropriation and inversion of Protagoras, from "man is the measure" to the "mismeasure of man." Gould was insinuating something Protagoras may or may not have intended by his famous dictum—that man, *as* the measure, was likely to misrepresent the truth about his object(s) of inquiry. For Gould, if not for Protagoras, these misrepresentations can be especially pernicious when the object of the inquiry is Man himself. My book too has been concerned with problems of human measurement, with what happens, and why, when Shakespeare's dramatic characters and poetic personae undertake to evaluate, assess, and appraise one another by their own "rules." If, however, Shakespeare's writing shares with the disciplines of mathematics or other quantifying sciences a preoccupation with measurement, it is not because Shakespeare means to conflate Renaissance literature and early modern science as a single intellectual or ethical field. Rather, he attempts to make a difference: Shakespeare's experiments with measurement demonstrate that there is no standard, no rule that can fully take the dimensions of human experience—or the realm of the imaginary.

Gould's subject, specifically, was the rise of theories and methods for the measurement of human intelligence, from early nineteenth-century craniology (measuring heads) to the development of the Intelligence Quotient (IQ) test (measuring minds). Both were driven by a

---

[1] Stephen Jay Gould, *The Mismeasure of Man*, rev. ed. (New York: W. W. Norton, 1996).

belief in a "measurable, genetically fixed, and unitary intelligence" and
the possibility of "its quantification as one number for each individual";
the purpose of the latter would be our capacity "to rank people in a
single series of worthiness."[2] As Gould's investigation makes clear, the
notion of establishing a single criterion for judging and comparing intelligence is a modern phenomenon. In the Renaissance, there was no
"test" to determine how smart one was in relation to others, no unitary,
quantified measure for rating comparative degrees of intellectual acuity
or strength. As modern readers and as scholars, accustomed to such
evaluations, we may be inclined to form opinions as to whether Hamlet
is more intelligent than Claudius, Falstaff than Hal, or Rosalind than
Orlando, on the basis of our own, implicit standards for assessment. But
would Shakespeare have dreamed of comparing his characters in such
terms? Although we often seem to consider Hamlet's distinctiveness as a
character, especially, according to the nature and degree of his intelligence, would Shakespeare have been likely, or able, to ask or answer the
question, How smart is Hamlet? In this brief conclusion, I would like to
speculate on the possibilities for "intelligence testing" in an age before
there was any single criterion for making this kind of judgment. In this,
I hope to come full circle in my study of the Renaissance *homo
mensura*—from the man who measures to the measure of a man's own
mind.

Renaissance views of the nature of the human mind are too many and
too complex to rehearse here. Whatever its particular configurations,
the early modern mind tends to be structured in terms of ancient faculty
psychology, with its attribution of different "powers" or functions, each
working separately or collaboratively to produce some aspect of perception and cognition. Often these faculties are organized hierarchically—
with some version of a "rational" faculty deemed "highest"—in a great
chain of mental being. These theories posit no single entity analogous to
our concept of "intelligence" but rather a set of "multiple intelligences"
operating in a single mind. Alternative models, such as the notion that
our minds incorporate "five wits" (sometimes identified as memory, fan-
tasy, judgment, imagination, and common wit), are also based on a di-
versification or compartmentalization of cognitive function.

Such theories tend to emphasize the nature of human minds generally, rather than provide a basis for comparing them. Nevertheless, fac-

---

[2] Ibid., 21.

ulty theory allows for differentiation in the same way that Renaissance humoral theory explains varying human temperaments, in terms of balance, excess, or deficiency, that is, the proportionality among its parts. Competing early modern theories, however, postulate divergent "kinds" of minds rather than variant distributions of elements within them. Roger Ascham, in his *Scholemaster*, famously distinguished between what he called "quick" and "hard" wits. Assigned by nature, but susceptible to cultivation, these two kinds of intelligence are not equal in capacity. Ascham assigns them new values: although earlier educators favored quick wits, Ascham makes the case that these are too "Headie, and Brainsicke" to produce good scholars.[3] Juan Huarte, in his *Examination of Mens Wits* (1594), expands the range of mental "kinds," offering the period's most systematic theory of "multiple intelligences." (And he knows it: he vaunts that no one before him "hath cleerely and distinctly deliuered what that nature is which maketh a man able for one science, and vncapable of another, nor how many differences of wittes there are found in mankind.")[4] Huarte insists that all brains are created equal: "[W]hen God formed Adam and Eue, it is certaine that before he filled them with wisdome, he instrumentalized their braine in such sort, as they might receiue it with ease, and serve as commodious instrument." However, whatever religion teaches us about human equality, he explains, it is clear that we are not equal in intellectual capacity: "But all soules being of equall perfection (as well that of the wiser, and that of the foolish) it cannot be affirmed, that nature in this signification, is that which makes a man able, for if this were true, all men should haue a like measure of wit and wisdome." Huarte invokes humoral theory to explain how men can have "equal" natures (in that they consist of the same four humors) but different "wits" (in that they possess those humors in different proportions). Along with his biochemical approach to intellectual diversity, Huarte also practices a kind of early modern craniology, correlating mental capacity with the shape and extent of the head: "Hence it followes that the man who hath his forehead very plaine, and his nodocke flat, hath not his braine so figured, as is requisit for wit and habilitie."[5] Ben Jonson

---

[3] Roger Ascham, *The Scholemaster*, in *Roger Ascham: The English Works*, ed. William Aldis Wright (Cambridge: Cambridge University Press, 1904), 188, 189.

[4] Juan Huarte, *The Examination of Mens Wits*, Trans. *M. Camillio Camilli, Englished by Richard Carew* (London, 1594), facsimile ed. (Gainesville, FL: Scholar's Facsimiles and Reprints, 1959), sig. A4v.

[5] Ibid., sigs. C3v, C5r.

makes an analogous *ingeniorum discrimina* in his *Timber, or Discoveries Made upon Men and Matter* (1641), arguing that "[t]here are no fewer forms of minds, than of bodies amongst us."[6]

Shakespeare seems to concur with his contemporaries in understanding the mind, in part, in qualitative, material terms. His lexicon, for example, includes many words and phrases (often, terms of abuse) that pertain to the physical nature of the "stupid" mind, such as "brainless," "clay-brained," "dim-witted," "dull," "dull-brained," "fat-brained," "fat-witted," and "idle-headed." The playwright imagines a "hot brain" or a "seething brain" as unusually inventive, while a "dry brain" retains memories but produces no new thoughts. Shakespeare also invokes the idea of the mind as a collection of wits and the corollary notion that these may be counted and compared. Beatrice mocks Benedick, for example, by remarking how "[i]n our last conflict four of his five wits went halting off, and now is the whole man govern'd with one" (*Much Ado about Nothing*, 1.1.65–67). The plays are filled with scenes of characters "matching" wits, competing to see who has "more" of them. I don't think that anyone, then or now, would disagree that Touchstone outwits William, Hamlet outwits Rosencranz and Guildenstern, or that Beatrice and Benedick "equal" each other, wit for wit. Shakespeare often enacts the measurement of wit through dramatic repartee, dialogue consisting of a point-for-point exchange of quips until the superior wit, quite literally, has the last word (i.e., one witticism *more* than his or her competition).

These examples are enough to indicate at least this much about "intelligence testing" in the Renaissance: in an age before there were explicit, agreed-upon standards or scales for evaluating intellectual power, there was already a range of ways of comparing minds with one another and valuing them accordingly. Although I have suggested that there was, as yet, no single, totalizing notion of "intelligence," one of its Renaissance versions comes closest, for Shakespeare, to representing an integrated mental excellence. I have implicitly argued throughout this book that Shakespeare considered judgment the "highest" form of intelligence, the kind of understanding ultimately sought by Brutus, Bertram, Hamlet, Othello, Leontes, and Lear. Judgment, as a faculty of early modern psychology and a goal of humanist education, something at once *in* the mind but also developed *by* the mind over time, has its own, long, and

---

[6] Ben Jonson, *Timber: or Discoveries*, in *Ben Jonson: The Complete Poems*, ed. George Parfitt (London: Penguin, 1988), 395.

complicated history. One of its key points of origin, Aristotle's discussion of "prudence" in the *Nichomachean Ethics*, established the man of judgment and understanding as the "standard and measure" of what is right and true.[7] Throughout the history of the concept, judgment has been identified with discrimination, comparison, proper proportioning, and evaluation.[8]

Edmund Spenser is among those who imagined it as a physical part of the brain, in his allegorical House of Alma, in book 2 of the *Faerie Queene*. Left mysteriously unnamed alongside Phantastes (Fancy or Imagination) and Eumnestes (Memory), Spenser's "judgment" is a "wise, and wondrous sage" who meditates on "all artes, all science, all Philosophy / And all that in the world was aye thought wittily."[9] For Spenser, judgment surveys all that is known; it is thought about *thought*. Judgment is continually "tested" in Renaissance literature—we may think of Spenser's knights, in their quests toward right understanding—but for Shakespeare, I have suggested, it is a test that proves almost too hard to pass. Whatever his views about judgment as an innate faculty of the mind, Shakespeare emphasizes its dependence on age and experience; yet, often enough, his protagonists arrive at it too late. Indeed, Shakespearean tragedy may be defined as the failure to fulfill the expectation that a fuller understanding comes to us in time, the broken promise of a final, perfected, *human* judgment. Although Spenser was able to "place" it, judgment, of all the Renaissance intelligences, is no doubt the most indeterminate of faculties as far as locating it is concerned (even Spenser could not name it); that is because it is also the "faculty of faculties," arbitrating across and among the mind's diverse operations. We may feel we know it when we see it in Shakespeare's poems and plays, yet it resists definition, defies absolute certainty. Of all the Renaissance human intelligences, it is the one that best represents the orchestration of the full and varied powers of the mind. It is the coordination and culmination of the *mens-mensura*, the mind-as-measure—and yet knows no measure itself.

In *All's Well That Ends Well*, a convocation of concerned gentlemen ex-

[7] For an excellent discussion of this passage and the notion of "prudence" in the Renaissance generally, see Victoria Kahn, *Rhetoric, Prudence, and Skepticism in the Renaissance* (Ithaca: Cornell University Press, 1985), 32ff.

[8] For a brilliant and comprehensive examination of the idea of "judgment" and the Renaissance arts, see David Summers, *The Judgment of Sense: Renaissance Naturalism and the Rise of Aesthetics* (Cambridge: Cambridge University Press, 1987).

[9] Edmund Spenser, *The Faerie Queene*, ed. Thomas P. Roche Jr. (New Haven: Yale University Press, 1978), 2.9.54.5; 2.9.53.8–9.

presses the wish that the misguided Bertram, who has grievously misjudged the value of the poor woman who loves him, "might take a measure of his own judgments" (4.3.33).[10] "True" judgment demands reflex-
ivity, a valid and accurate measure of the measure (as the earlier chapters
in this book have shown). As modern readers of Shakespeare, we often
value any attempt at reflexiveness and rate self-consciousness and self-
scrutiny as markers of high intelligence in his dramatic characters. By
the standard of "self-consciousness," to be sure, characters such as Brutus
and Hamlet are "gifted," set apart, as they often are, from their unre-
flecting peers. Yet however admirable Shakespeare might have found the
act in its own right, self-scrutiny does not necessarily imply self-
understanding in his poems and plays. Too often, even the most discerning judge is compromised in his own case.

How smart, then, is Shakespeare's Hamlet? Faculty by faculty, his may
be the foremost intelligence of Shakespeare's characters. He is very witty.
He has a vivid imagination and an unyielding memory. He reasons ex-
ceptionally well. And he is surpassingly cunning. Yet for all of Hamlet's
prodigious intelligences, he may not actually be, in Shakespeare's terms,
all that "smart." He has a bad habit of passing judgment on others far in
excess of their faults, for example, in his cruel treatment of Ophelia
and his precipitous murder of Polonius. For all his efforts at self-
understanding, he is often tragically uncertain of his own motives, his
own desires. Lacking a degree of human judgment, Hamlet's mind ex-
pends its powers and leaves him, often enough, imperceptive and un-
awares.

When Hamlet rhapsodizes on Man as "infinite in faculties" (2.2.304),
he gives voice (albeit, in his case, with some irony) to the optimistic
strain of Renaissance humanism, the one that heard "man is the measure" as asserting the potentially limitless scope of human apprehension.
For Renaissance writers of a more skeptical bent, a mind "beyond measure" is a mind that, in effect, *can't* be measured or counted on to know
with any certitude. With Seneca, Shakespeare marveled how the *homo
mensura*, for all his accomplishments in measuring the world around
him, cannot seem to compass himself:

---

[10] Bertram's Helen of *All's Well* is explicitly compared to Helen of Troy ("'Was this fair
face the cause,' quoth she, 'Why the Grecians sacked Troy?'" [1.3.70–71]); once
again, as in *Troilus and Cressida*, Helen seems to be a key site for Shakespeare's debates
on human value.

O what marvelous skill! You know how to measure the circle; you find the square of any shape which is set before you; you compute the distances between the stars; there is nothing which does not come within the scope of your calculations. But if you are a real master of your profession, measure me the mind of man! Tell me how great it is, or how puny! You know what a straight line is; but how does it benefit you if you do not know what is straight in this life of yours![11]

On the question of human measurement, Shakespeare too comes full circle: the mind that measures cannot measure itself. For him, the truth of who we are and the extent of what we know awaits the final Judgment, the only certain Reckoning. The mind's true measure is beyond our capacity, beyond our place and time. Yet even so, for Shakespeare, what's left to us is still worth knowing—the range of the rules we apply, the contrariety of measures we mete, in our ceaseless efforts to find it.

[11] Lucius Annaeus Seneca, *Epistles 66–93*, trans. R. M. Gummere (Cambridge: Harvard University Press, 1992), vol. 5, epistle 88.

BIBLIOGRAPHY

Adelman, Janet. "'Her Father's Blood': Race, Conversion, and Nation in the *Merchant of Venice*." *Representations* 81 (Winter 2003): 4–30.
——. "Iago's Alter-Ego: Race as Projection in *Othello*." *Shakespeare Quarterly* 48.2 (1997): 125–44.
Africanus, Leo. *The History and Description of Africa*. 3 vols. Trans. John Pory (1600), ed. Robert Brown. New York: Burt Franklin, 1963.
Alberti, Leon Battista. *Libri della famiglia*. Trans. Renée Neu Watkins. Columbia: University of South Carolina Press, 1969.
Alford, John A. "Literature and Law in Medieval England." *PMLA* 92 (1977): 941–51.
Aquinas, Thomas. *Summa theologiae*. In *St. Thomas Aquinas: Political Writings*. Ed. and trans. R. W. Dyson. Cambridge: Cambridge University Press, 2002.
——. *Truth*. 3 vols. Trans. Robert W. Mulligan. Indianapolis: Hackett Publishing, 1954.
Aristotle. *Metaphysics*. Trans. Hugh Tredennik. Loeb Classical Library. London: William Heinemann, 1933.
——. *Nichomachean Ethics*. Trans. H. Rackham. Loeb Classical Library. London: William Heinemann, 1947.
Ascham, Roger. *The Scholemaster*. In *Roger Ascham: The English Works*, ed. William Aldis Wright. Cambridge: Cambridge University Press, 1904.
Ashley, Robert, trans. *A Comparison of the English and Spanish Nation*. London, 1589.
Attridge, Derek. *The Rhythms of English Poetry*. New York: Longman, 1982.
——. *Well-Weigh'd Syllables: Elizabethan Verse in Classical Metres*. Cambridge: Cambridge University Press, 1974.
Bacon, Francis. *The Advancement of Learning*. In *The Works of Francis Bacon*, vol. 5., ed. James Spedding, Robert Leslie Ellis, and Douglas Denon Heath. New York: Garret Press, 1968.
——. *The New Organon*. Ed. Lisa Jardine and Michael Silverthorne. Cambridge: Cambridge University Press, 2000.
Bady, David. "The Sum of Something: Arithmetic in *The Merchant of Venice*." *Shakespeare Quarterly* 36.1 (1985): 10–30.
Baker, G. P., and P. M. S. Hacker. "The Standard Metre." In *Wittgenstein: Understanding and Meaning*, ed. G. P. Baker and P. M. S. Baker. Oxford: Blackwell, 1980.
Barfoot, C. C. "*Troilus and Cressida*: 'Praise us as we are tasted.'" *Shakespeare Quarterly* 39.1 (Spring 1988): 45–57.
Berka, Karel. *Measurement: Its Concepts, Theories, and Problems*. Trans. Augustin Riska. Boston Studies in the Philosophy of Science 72. Dordrecht, Holland: D. Reidel, 1983.
Berriman, A. E. *Historical Metrology*. London: Dent and Sons, 1953.

Blackstone, William T., ed. *The Concept of Equality*. Minneapolis: Burgess, 1969.

Bodin, Jean. *The Sixe Bookes of the Commonweale*. Facsimile ed. Ed. Kenneth Douglas McRae. Cambridge, MA: Harvard University Press, 1962.

A Booke Conteining All Svch Proclamations, as were published dvring the Raigne of the late Queene Elizabeth. London, 1618.

The Book of Common Prayer, 1559: The Elizabethan Prayer Book. Ed. John E. Booty. Washington, DC: Folger Shakespeare Library, 1976.

Booth, Stephen, ed. *Shakespeare's Sonnets*. New Haven: Yale University Press, 1977.

Bridgman, P. W. *The Way Things Are*. Cambridge: Cambridge University Press, 1959.

Brooke, Tucker. "Marlowe's Versification and Style." *Studies in Philology* 19 (1922): 186–205.

Burkhardt, Sigurd. *Shakespearean Meanings*. Princeton: Princeton University Press, 1968.

Carlyle, R. W., and A. J. Carlyle. *A History of Medieval Political Theory in the West*. 6 vols. Edinburgh: William Blackwood, 1903–36.

Carnap, Rudolf. *An Introduction to the Philosophy of Science*. Ed. Martin Gardner. New York: Basic Books, 1966.

Chambers, E. K. *William Shakespeare: A Study of Facts and Problems*. Oxford: Clarendon Press, 1930.

Cicero. *De legibus*. In *De re publica; De legibus*. Trans. Clinton Walker Keyes. Loeb Classical Library. London: William Heinemann, 1928.

Cinthio, Giraldi. *The Moor of Venice*. Trans. John Edward Taylor. New York: Ams Press, 1972.

Cole, Thomas A. "Classical Greek and Latin." In *Versification: Major Language Types*, ed. W. K. Wimsatt, 66–88. New York: New York University Press, 1972.

Connor, R. D. *The Weights and Measures of England*. London: Her Majesty's Stationery Office, 1987.

Colyn, John. *Exposure of the Abuses of Weights and Measures for the Information of the Council*. London, ca. 1517.

Contareno, Gaspar. *The Commonwealth and Gouernment of Venice*. London, 1599.

Crane, Mary Thomas. *Shakespeare's Brain: Reading with Cognitive Theory*. Princeton: Princeton University Press, 2001.

Crosby, Alfred W. *The Measure of Reality: Quantification and Western Society, 1250–1600*. Cambridge: Cambridge University Press, 1997.

Costalius, Petrus [Pierre Coustau]. *Pegma*. Lugdini: Apud Matthiam Bonhomme, 1555.

Culpeper, Nicholas, trans. *The Anatomy of the Body of Man*. London, 1653.

Cusa, Nicholas de. *Idiota de Mente*. Trans. Clyde Lee Miller. New York: Abaris Books, 1979.

Cutler, Allan Harris, and Helen Elmquist Cutler. *The Jew as Ally of the Muslim: Medieval Roots of Anti-Semitism*. Notre Dame: University of Notre Dame Press, 1986.

Day, John, William Rowley, and George Wilkins. *The Travels of the Three English Brothers*. In *Three Renaissance Travel Plays*, ed. Anthony Parr. Manchester: Manchester University Press, 1995.

Dear, Peter. *Discipline and Experience: The Mathematical Way in the Scientific Revolution*. Chicago: University of Chicago Press, 1995.

Dee, John. "Mathematical Preface." In *Euclid*, trans. Henry Billingsley. London, 1570.

Dent, J. Geoffrey. "The Pound Weight and the Pound Sterling." *Folk Life* 27 (1988–89): 80–84.

Diamond, Cora. "How Long Is the Standard Meter in Paris?" In *Wittgenstein in America*, ed. T. G. McCarthy and S. C. Stidd, 104–39. Oxford: Oxford University Press, 2001.

Dingle, Herbert. "A Theory of Measurement." *British Journal for the Philosophy of Science* 1.1 (May 1950): 5–26.

Dollimore, Jonathan. *Radical Tragedy: Religion, Ideology and Power in the Drama of Shakespeare and His Contemporaries*. Chicago: Chicago University Press, 1984.

Eden, Kathy. *Poetic and Legal Fiction in the Aristotelian Tradition*. Princeton: Princeton University Press, 1986.

Elyot, Sir Thomas. *The Booke Named the Governor*. Facsimile ed. Menston: Scolar Press, 1970.

——. *The Dictionary*. London, n.d.

Elton, William. *King Lear and the Gods*. Lexington: University Press of Kentucky, 1988.

Erasmus, Desiderius. *Adagia*. In *The Collected Works of Erasmus*, vol. 31. Toronto: University of Toronto Press, 1982.

Everett, Barbara. "'Spanish' Othello: The Making of Shakespeare's Moor." *Shakespeare Survey* 35 (1982): 101–12.

Ferry, Anne. *The Inward Language: Sonnets of Wyatt, Sidney, Shakespeare, Donne*. Chicago: Chicago University Press, 1983.

Fineman, Joel. *Shakespeare's Perjur'd Eye: The Invention of Poetic Subjectivity in the Sonnets*. Berkeley: University of California Press.

Fischer, Sandra K. *Econolingua: A Glossary of Coins and Economic Language in Renaissance Drama*. Newark: University of Delaware Press, 1985.

Fletcher, Giles. *Christs Victorie in Heaven*. Cambridge, 1610.

Fletcher, John, and Philip Massinger. *The Custom of the Country*. Ed. Nick de Somogyi. Cambridge: Routledge, 1999.

Florio, John. *Queen Anna's new world of words*. London, 1611.

——. *A worlde of wordes*. London, 1598.

Fowler, Alastair. *Spenser and the Numbers of Time*. New York: Barnes and Noble, 1964.

——. *Triumphal Forms: Structural Patterns in Elizabethan Poetry*. Cambridge: Cambridge University Press, 1970.

Fowler, Elizabeth. "The Failure of Moral Philosophy in the Work of Spenser." *Representations* 51 (Summer 1995): 47–76.

Fraunce, Abraham. *The Lawiers Logike*. London, 1588.

Galen. *Certain workes of Galens, called Methodus medendi*. Trans. T. Gale. London, 1585.

——. *The Diagnosis and Cure of the Soul's Passions*. In *Galen on the Passions and Errors of the Soul*, trans. Paul W. Harkins. Columbus: Ohio State University Press, 1963.

The Geneva Bible: Facsimile Edition of 1560. Ed. Lloyd E. Berry. Madison: University of Wisconsin Press, 1969.

Geveren, Scheltco. *Of the ende of this worlde, and seconde commynge of Christe.* London, 1578.

Gould, Stephen Jay. *The Mismeasure of Man.* Rev. ed. New York: W. W. Norton, 1996.

Greaves, Paul. *Discourse of the Roman Foot.* London, 1647.

Griffin, Eric. "Un-Sainting James: Or, Othello and the 'Spanish Spirits' of Shakespeare's Globe." *Representations* 62 (Spring 1998): 58–99.

Hadfield, Andrew. "Race in *Othello: The Description of Africa* and the Black Legend." *Notes & Queries* 45 (September 1998): 336–38.

Hall, Hubert, ed. *Select Tracts and Table Books Relating to English Weights and Measures, 1100–1742.* Camden Miscellany 15. London: Offices of the Society, 1929.

Halpern, Richard. *The Poetics of Primitive Accumulation: English Renaissance Culture and the Genealogy of Capital.* Ithaca: Cornell University Press, 1991.

Hanmer, Meredith. *The Baptizing of a Turke.* London, 1586.

Hannaford, Ivan. *Race: The History of an Idea in the West.* Baltimore: Johns Hopkins University Press, 1996.

Hanson, Kristin. "From Dante to Pinsky: A Theoretical Perspective on the History of the Modern English Iambic Pentameter." *Rivista di Linguistica* 9.1 (1996): 53–97.

Hardison, O. B. *Prosody and Purpose in the English Renaissance.* Baltimore: Johns Hopkins University Press, 1989.

Harris, Jonathan Gil. *Foreign Bodies and the Body Politic: Discourses of Social Pathology in Early Modern England.* Cambridge: Cambridge University Press, 1998.

Hart, James S., Jr. *The Rule of Law, 1603–1660: Crowns, Courts, and Judges.* Harlow: Pearson Longman, 2003.

Head, Randolph C. "Religious Boundaries and the Inquisition in Venice: Trials of Jews and Judaizers, 1548–1580." *Journal of Medieval and Renaissance Studies* 20.2 (1990): 175–204.

Hess, Andrew C. "The Moriscos: An Ottoman Fifth Column in Sixteenth-Century Spain." *American Historical Review* 74 (1968): 1–25.

Hill, Thomas. *The arte of vulgare arithmeticke.* London, 1600.

Hobbes, Thomas. *Dialogue between a Philosopher and a Student of the Common Laws of England.* In *The Collected Works of Thomas Hobbes,* ed. Sir William Mollesworth, vol. 6. Bristol: Thoemmes Press, 1994.

Hooker, Richard. *Of the Laws of Ecclesiastical Polity.* In *The Works of Mr. Richard Hooker,* ed. Rev. John Keble. Oxford: Clarendon Press, 1888.

Howson, Geoffrey. *A History of Mathematics Education in England.* Cambridge: Cambridge University Press, 1982.

Huarte, Juan. *The Examination of Mens Wits, Trans. M. Camillio Camilli, Englished by Richard Carew.* London, 1594. Facsimile ed. Gainesville, FL: Scholar's Facsimiles and Reprints, 1959.

Jaffee, Michele Sharon. *The Story of O: Prostitutes and Other Good-for-Nothings in the Renaissance.* Cambridge, MA: Harvard University Press, 1999.

Jones, Emrys. "*Othello,* Lepanto and the Cyprus Wars." *Shakespeare Survey* 21 (1968): 47–52.

Jonson, Ben. *Timber, or Discoveries.* In *Ben Jonson: The Complete Poems,* ed. George Parfitt. London: Penguin, 1988.

Kahn, Victoria. *Rhetoric, Prudence, and Skepticism in the Renaissance.* Ithaca: Cornell University Press, 1985.

Kaye, Joel. *Economy and Nature in the Fourteenth Century: Money, Market Exchange, and the Emergence of Scientific Thought.* Cambridge: Cambridge University Press, 1998.

Kamen, Henry. *The Spanish Inquisition: A Historical Revision.* New Haven: Yale University Press, 1997.

Kiparsky, Paul. "The Rhythms of English Verse." *Linguistic Inquiry* 8 (1977): 189–247.

Kline, Morris. *Mathematics in Western Culture.* Oxford: Oxford University Press, 1964.

Kronenfeld, Judith. *King Lear and the Naked Truth: Rethinking the Language of Religion and Resistance.* Durham, NC: Duke University Press, 1998.

Kula, Witold. *Measures and Men.* Trans. R. Szreter. Princeton: Princeton University Press, 1986.

Lakoff, George, and Rafael E. Nunez. *Where Mathematics Comes From: How the Embodied Mind Brings Mathematics into Being.* New York: Basic Books, 2000.

Laquer, Thomas. *Making Sex: Body and Gender from the Greeks to Freud.* Cambridge, MA: Harvard University Press, 1990.

Leonardo da Vinci. *The Notebooks of Leonardo da Vinci.* 2 vols. Ed. Jean Paul Richter. New York: Dover Publications, 1883.

Lewkenor, Sir Lewis. *Discourse of the Usage of the English Fugitives by the Spaniard.* London, 1595.

Loomba, Ania. *Shakespeare, Race, and Colonialism.* Oxford: Oxford University Press, 2002.

Lupton, Julia Reinhard. "Othello Circumcised: Shakespeare and the Pauline Discourse of Nations." *Representations* 57 (Winter 1997): 73–89.

Lust's Dominion or The Lascivious Queen. Ed. Khalid Bekkaoui. Fez, 1999.

Machiavelli, Niccolò. *Florentine Histories.* Trans. Laura F. Banfield and Harvey C. Mansfield Jr. Princeton: Princeton University Press, 1988.

Maclean, Ian. *Interpretation and Meaning in the Renaissance: The Case of Law.* Cambridge: Cambridge University Press, 1992.

Maltby, William S. *The Black Legend in England: The Development of Anti-Spanish Sentiment, 1558–1660.* Durham, NC: Duke University Press, 1971.

Marlowe, Christopher. *The Complete Works of Christopher Marlowe.* 2 vols. Ed. Fredson Bowers. Cambridge: Cambridge University Press, 1973.

Marston, John. *Antonio's Revenge.* Ed. Reavely Gair. Manchester: Manchester University Press, 1999.

Masterson, Thomas. *First Booke of Arithmeticke.* London, 1592.

Menninger, Karl. *Number Words and Number Symbols: A Cultural History of Numbers.* Trans. Paul Broneer. Cambridge, MA: MIT Press, 1969.

Milton, John. *Paradise Lost.* In *John Milton: Complete Poems and Major Prose,* ed. Merritt Y. Hughes. New York: Macmillan, 1957.

Montaigne, Michel de. *The Complete Essays of Montaigne.* Trans. Donald M. Frame. Stanford: Stanford University Press, 1957.

Moore, Peter. "Shakespeare's Iago and Santiago Matamoros." *Notes & Queries* 43 (June 1996): 162–63.

Munday, Anthony, trans. *The knowledge of a mans owne selfe by P. De Mornay.* London, 1602.

Murdoch, John E. "From Social into Intellectual Factors: An Aspect of the Unitary Character of Medieval Learning." In *The Cultural Context of Medieval Learning,* ed. John E. Murdoch and Edith Sylla, 271–348. Dordrecht, Holland: D. Reidel, 1975.

Neill, Michael. "'Mulattos,' 'Blacks,' and 'Indian Moors': *Othello* and Early Modern Constructions of Human Difference." *Shakespeare Quarterly* 49.2 (1998): 361–74.

Newman, James R. *The World of Mathematics.* New York: Simon and Schuster, 1956.

Nicholson, E. *Men and Measures.* London: Smith and Elder, 1912.

Parker, Patricia. *Shakespeare from the Margins: Language, Culture, Context.* Chicago: University of Chicago Press, 1996.

Patterson, Annabel. "The Egalitarian Giant: Representations of Justice in History/Literature, *Journal of British Studies* 31 (April 1992): 97–132.

Penkethman, John. *Artachthos, or, a New Booke declaring the Assize or Weight of Bread.* London, 1638.

Pennock, J. Roland, and John W. Chapman, eds. *Equality.* New York: Atherton Press, 1967.

Perez, Antonio. *A Treatise Paranaenetical.* London, 1598.

Pintore, Anna. *Law without Truth.* Liverpool: Deborah Charles Publications, 2000.

Plato. *Laws.* 2 vols. Trans. R. G. Bury. Loeb Classical Library. Cambridge, MA: Harvard University Press, 1926.

———. *Theaetetus.* Trans. John McDowell. Oxford: Clarendon Press, 1973.

Plowden, Edward. *An Exact Abridgment in English of the Commentaries, or Reports of the learned and famous Lawyer, Edmond Plowden.* London, 1650.

Pont, Richard. *New treatise of the right reckoning of yeares.* Edinburgh, 1599.

Poole, Matthew. *Annotations upon the Holy Bible.* London, 1683.

Popkin, Richard. *The History of Scepticism from Erasmus to Spinoza.* Berkeley: University of California Press, 1979.

Pullan, Brian. *The Jews of Europe and the Inquisition of Venice, 1550–1670.* Totowa, NJ: Barnes and Noble, 1983.

Puttenham, George. *The Arte of English Poesie.* Ed. Gladys Doidge Willcock and Alice Walker. Cambridge: Cambridge University Press, 1970.

Ramsay, Paul. "The Syllables of Shakespeare's Sonnets." In *New Essays on Shakespeare's Sonnets,* ed. Hilton Landry, 193–215. New York: AMS Press, 1976.

Rea, John A. "Iago." *Names: A Journal of Onomastics* 34.1 (March 1986): 97–98.

Recorde, Robert. *Ground of Artes.* London, 1543.

———. *Whetstone of Witte.* London, 1557.

Rotman, Brian. *Signifying Nothing.* New York: St. Martin's Press, 1987.

R.W. *The Three Ladies of London* (1584). Ed. H. S. D. Mithal. New York: Garland, 1988.

Sackville, Thomas. "Amplissimo." Preface to Castiglione's *The Courtier.* London, 1577.

Scarry, Elaine. *On Beauty and Being Just.* Princeton: Princeton University Press, 1999.

Scodel, Joshua. *Excess and the Mean in Early Modern English Literature.* Princeton: Princeton University Press, 2002.

Scott, Izora. *Controversies over the Imitation of Cicero in the Renaissance.* Davis, CA: Hermagoras Press, 1991.

Selden, John. *The Table-Talk of John Selden.* London: William Pickering, 1847.

Seneca, Lucius Annaeus. *Epistulae Morales.* Trans. R. M. Gummere. In *Seneca in Ten Volumes.* Loeb Classical Library. Cambridge, MA: Harvard University Press, 1971.

Sextus Empiricus. *Against the Logicians.* Trans. R. G. Bury. Loeb Classical Library. London: William Heinemann, 1976.

Shakespeare, William. *The Riverside Shakespeare.* 2nd ed. Ed. G. Blakemore Evans. Boston: Houghton Mifflin, 1997.

Shannon, Laurie. "Nature's Bias: Renaissance Homonormativity and Elizabethan Comic Likeness." *Modern Philology* 98.2 (November 2000): 183–210.

———. *Sovereign Amity: Figures of Friendship in Shakespearean Contexts.* Chicago: Chicago University Press.

Shapiro, James. *Shakespeare and the Jews.* New York: Columbia University Press, 1996.

Siraisi, Nancy G. "Vesalius and Human Diversity in *De humani corporis fabrica.*" *Journal of the Warburg and Courtauld Institutes* 57 (1994): 60–88.

Smith, Barbara Hernnstein. *Contingencies of Value: Alternative Perspectives for Critical Theory.* Cambridge, MA: Harvard University Press, 1988.

Smith, G. Gregory, ed. *Elizabethan Critical Essays.* 2 vols. Oxford: Oxford University Press, 1904.

Spenser, Edmund. *The Faerie Queene.* Ed. Thomas P. Roche Jr. New York: Penguin, 1978.

Spiller, Elizabeth. *Science, Reading, and Renaissance Literature.* Cambridge: Cambridge University Press, 2004.

Summers, David. *The Judgment of Sense: Renaissance Naturalism and the Rise of Aesthetics.* Cambridge: Cambridge University Press, 1987.

Tebbit, Mark. *Philosophy of Law: An Introduction.* London: Routledge, 2000.

Trinkhaus, Charles. "Protagoras in the Renaissance: An Exploration." In *Philosophy and Humanism: Renaissance Essays in Honor of Paul Oskar Kristeller,* ed. Edward P. Mahoney, 201–2. New York: Columbia University Press, 1976.

Tymme, Thomas. *The Court of Conscience.* London, 1605.

Vitkus, Daniel, ed. *Three Turk Plays from Early Modern England: "Selimus," "A Christian Turned Turk," and "The Renegado."* New York: Columbia University Press, 2000.

Westen, Peter. *Speaking of Equality: An Analysis of the Rhetorical Force of "Equality" in Moral and Legal Discourse.* Princeton: Princeton University Press, 1990.

White, R. S. *Natural Law in English Renaissance Literature.* Cambridge: Cambridge University Press, 1996.

Wilson, Mary Floyd. *English Ethnicity and Race in Early Modern Drama.* Cambridge: Cambridge University Press, 2003.

Wilson, Thomas. *A Discourse upon Usury.* G. Bell, 1925.

Wittgenstein, Ludwig. *The Blue and Brown Books.* 2nd ed. New York: Harper and Row, 1960.

———. *Philosophical Investigations.* Trans. G. E. M. Anscombe. Oxford: Basil Blackwell, 1967.

———. *Philosophical Remarks.* Ed. Rush Rhees. Trans. Raymond Hargreaves and Roger White. Oxford: Basil Blackwell, 1975.

Woodbridge, Linda, ed. *Money and the Age of Shakespeare: Essays in New Economic Criticism.* New York: Palgrave Macmillan, 2003.

Woods, Suzanne. *Natural Emphasis: English Versification from Chaucer to Dryden.* San Marino: Huntington Library, 1984.

Wright, George T. *Shakespeare's Metrical Art.* Berkeley: University of California Press, 1988.